REBECCA NASON

RSPB BIRDS OF THE
BRITISH ISLES

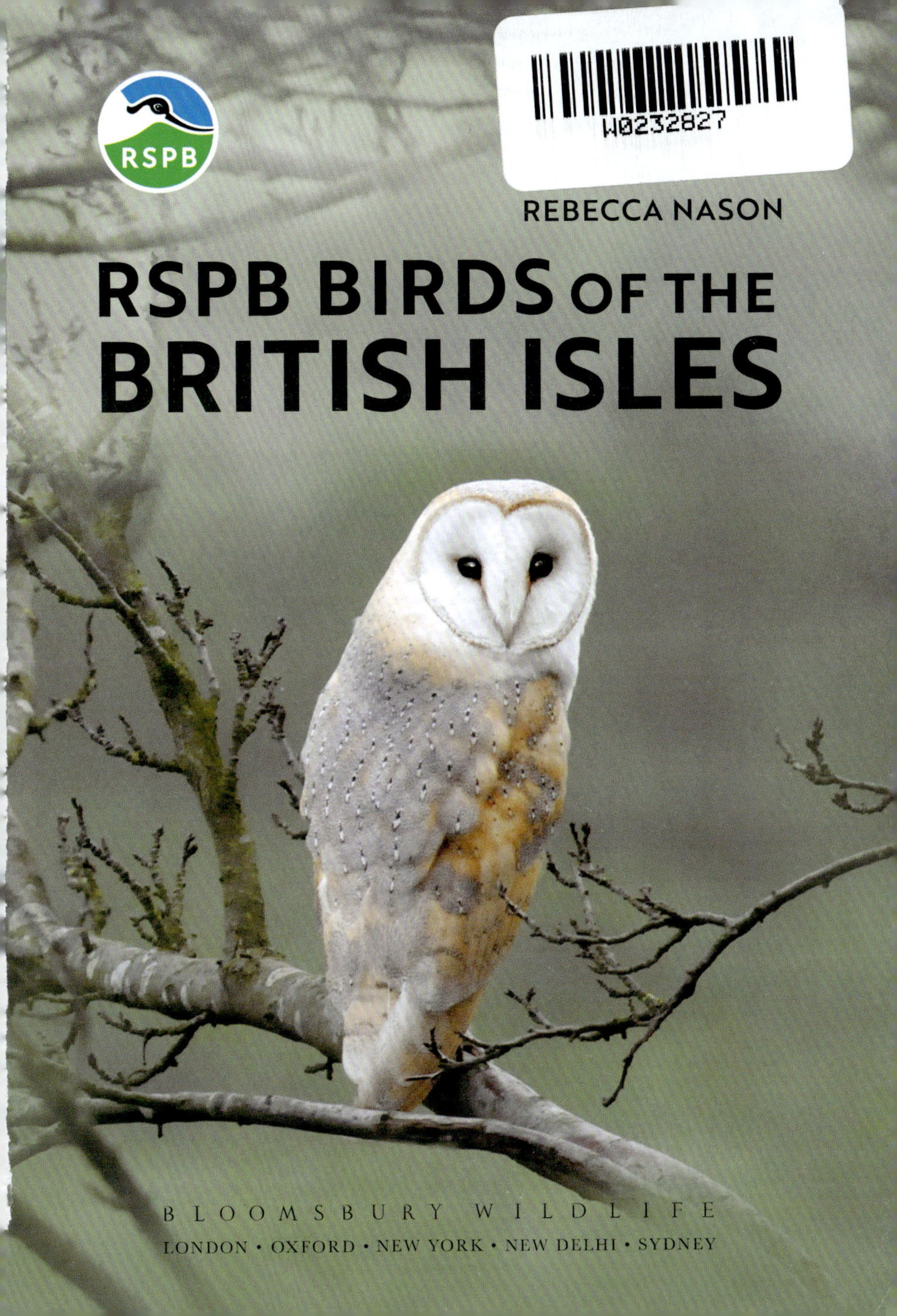

BLOOMSBURY WILDLIFE

LONDON · OXFORD · NEW YORK · NEW DELHI · SYDNEY

BLOOMSBURY WILDLIFE
Bloomsbury Publishing Plc
50 Bedford Square, London, WC1B 3DP, UK
29 Earlsfort Terrace, Dublin 2, Ireland

BLOOMSBURY, BLOOMSBURY WILDLIFE and the Diana logo are trademarks of
Bloomsbury Publishing Plc

First published in the United Kingdom 2024

A catalogue record for this book is available from the British Library.

ISBN: PB: 978-1-3994-0083-1
ePub: 978-1-3994-0082-4
ePDF: 978-1-3994-0081-7

2 4 6 8 10 9 7 5 3 1

Design by Susan McIntyre
Cover design by Amanda Keyte

Printed and bound in Turkey by Elma Basim

100%
From well-
managed forests
FSC   FSC® C164814
www.fsc.org

To find out more about our authors and books visit www.bloomsbury.com
and sign up for our newsletters.

Contents

Introduction   4

Species accounts   9

Glossary   268

Acknowledgements   268

Photo credits   269

Index   270

Introduction

Home to a diverse wealth of species and habitats, the British Isles are a fantastic place to be a birdwatcher. Across mountains, woodland, moorland, marshes, rivers, lakes and thousands of miles of coastline, birders can spot a brilliant array of species, and each year millions of birds travel to or through these islands on migration. Reflecting this variety, this guide covers 280 species you are likely to spot in Britain, Ireland and the Isle of Man, from common species to scarcer and rarer migratory birds.

Where to birdwatch

You don't need to travel far to birdwatch in the British Isles. Wherever you live – be it in coastal Ireland, inland rural Scotland or a big city in central England – there is a wealth of birdlife right on your doorstep. You might be surprised by what flies through your own garden or local park – a flock of Long-tailed Tits, a Sparrowhawk swooping between houses, or the electric blue flash of a Kingfisher visiting a pond.

Lesser Spotted Woodpecker (female)

The British Isles are home to a large number of nature reserves, 400 of which are designated National Nature Reserves. Some of these are run by the RSPB. There are 170 RSPB nature reserves that you can visit, and these reserves offer a fabulous array of birdwatching and wellbeing opportunities, each with amazing seasonal variation. They are brilliant for relaxed days 'out in the field' and perfect for any age, children and adults alike. With habitat and species protection at their core, nature reserves provide freedom to enjoy the great outdoors and develop birdwatching skills at your own pace. Many have hides set up at strategic locations with identification charts and aides, where you can watch woodland birds coming to nearby bird feeders, or set up a telescope and savour the delights of tideline waders. Some reserves offer extensive walks through beautiful open habitats, while smaller suburban sites feature child-orientated nature activities and feeding stations for birds and people alike. For a relatively small group of islands, we are spoilt with the range of habitats and birds around us, which are briefly described below.

Farmland covers three-quarters of the British Isles, and many birds thrive in this habitat and are considered 'farmland species'. Some are struggling to survive due to our current farming methods, and farmland bird numbers are in serious decline in many parts of the UK. Key species include Skylark and Lapwing, while mature hedgerows between fields are a vital habitat for Yellowhammer, House Sparrow and Linnet and offer shelter and food for winter visitors such as Redwing and Fieldfare.

Coastal habitats are hugely important, as cliffs offer safety from most predators and are excellent nesting locations for otherwise vulnerable breeding seabirds. In summer, some cliffs are alive with internationally important numbers of Gannet, Kittiwake, Puffin, Guillemot and Razorbill. Beaches are important feeding sites for Sanderling, Turnstone and Oystercatcher which feast on tideline invertebrates, while rocky shores are roosting sites for flocks of waders, gulls and terns.

Estuaries and marshlands are great places to see a variety of species all year round, as many resident and migratory birds feed on mud-dwelling invertebrates. Look out for Curlew, Oystercatcher, Bar-tailed Godwit, Ringed Plover, and maybe even Avocet and Grey Plover. Saltmarshes are alive with birds in autumn and winter, and here you can spot flocks of ducks, geese and waders, as well as uncommon species like Shore Lark and Snow Bunting. Spring is an excellent time to see pipits and wagtails on coastal grasslands.

Wetlands (rivers, streams, ponds, wet meadows and reedbeds) are havens for birdlife. Watch out for swans, geese, ducks and grebes on rivers and reservoirs; and herons, egrets, waders, pipits and wagtails along wetland fringes. Ponds, lakes and rivers may harbour other

Long-tailed Duck (male)

delights – a riverside Kingfisher, a dipper on a fast moving river or a Goosander on a winter lake. Wet meadows support many breeding and wintering waders, and can be excellent places to spot large birds of prey.

Meadows and grasslands, ranging from chalk grasslands to heathy margins, host an amazing range of wildlife. Wild grasslands have been dramatically reduced and fragmented since the 1930s, with more than 97% already lost. Grassland species include Skylark, Common Snipe, Green Woodpecker, Meadow Pipit, Barn Owl and Stone-curlew.

Heathlands are wide open, diverse habitats dominated by short plants such as gorse, heather, broom and grasses. They support a wealth of important species of plants and insects, which support a range of specialised bird species including Woodlark, Dartford Warbler, Hobby, Tree Pipit, Linnet, Nightjar and Stonechat.

Ptarmigan (winter male)

Mountains and upland moorlands are home to a distinct range of birds and many uplands are covered in peat, a vital natural store for carbon. Key species in upland mountain regions include Ptarmigan, Black Grouse, Ring Ouzel, Dotterel and Golden Eagle. Some upland moors support breeding waders such as Golden Plover, Redshank, Whimbrel and Curlew, and raptors such as Hen Harrier.

Towns and cities are home to many species that have adapted to urban habitats. Peregrine nest in many towns and cities, while feral pigeons dominate our inner-city spaces and several gull species thrive within coastal settlements. Urban lakes and rivers support Mallard, Coot, Little Grebe and Grey Heron, and large parks can be excellent places to spot Jay and Green Woodpecker as well as tit flocks in winter. Gardens are havens for Robin, Dunnock, Wren, Great Spotted Woodpecker, Goldfinch and Chaffinch, and some may even attract Treecreeper and Nuthatch.

Woodlands offer rich and diverse habitats, whether coniferous, deciduous, or mixed. Deciduous woodland can be home to Tawny Owl, and some woodlands support the scarce Lesser Spotted Woodpecker. More specific woodlands such as oak woodlands support Wood Warbler and Pied Flycatcher. Coniferous woodlands may be home to Crossbill, Siskin, Goldcrest and even Capercaillie in Old Scots pine forest.

Identifying birds

Being able to see, hear and name the species you are looking at is a wonderful feeling that connects you with the natural world. Differentiating between similar species can sometimes be a challenge, but this book provides everything you need to build your birding knowledge. There are several things that can help you to make the most out of any birdwatching experience:

- When looking at an individual bird, pay attention to its shape, colour, size, patterning, behaviour, habitat, its call or song, and your location and time of year.
- Consider purchasing binoculars, which are an essential aide for identification. You may also opt to use a telescope, which can give you closer views of distant subjects and allow for detailed study of a species while minimalising disturbance. A light-weight, water- or weather-proof camera is another excellent tool for birders, and can really help with identifying features both in the field and back home.

- Learning bird sounds isn't easy, but once you start to listen you open up a new dimension of bird identification. There are excellent online resources and apps that provide clear bird calls and songs.
- Keep fieldnotes on your phone or in a notebook to help remember places, dates, times and species seen. Note-taking can really help beginners on their birding journey, while experienced birders often write detailed notes and descriptions when reporting rare bird sightings.
- Be mindful of wildlife and other birdwatchers while in the field. It's important to be quiet so you don't disturb birds that are sensitive to sudden movement and noise, and you will be able to observe for longer.

Distribution map

The colour-coded distribution map covers the entire British Isles and provides a visual representation of where you are most likely to see the bird at different times of year.

Green: resident – areas where the species may be seen throughout the year and where it breeds.

Yellow: summer visitor – areas where the species may be seen in summer and usually breed.

Blue: winter visitor – areas where the species spends the winter, but does not breed.

Pink: passage migrant – areas that the species visits at times of migration, generally spring and autumn.

The more time spent out in the field, the more you will develop an understanding of different species in their environment and hone your own fieldcraft skills. Over time, an experienced birdwatcher can identify commonly seen birds from just a split-second view or on a call alone, while uncommon species may require more time to identify.

How to use this book

For each species, this book contains a concise description of the key identification features – shape, size, colour, patterning, behaviour, habitat and call or song. Each entry (except for a handful of very scarce birds) also includes a distribution map and a brief description of when and where to see each species.

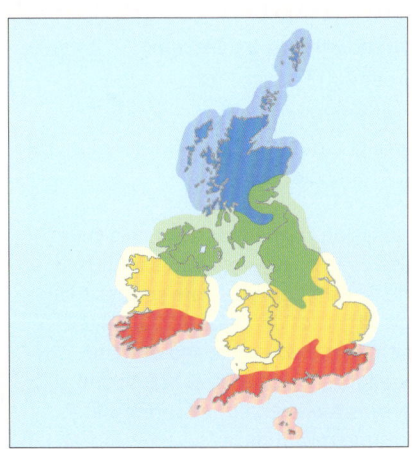

For distribution at sea, colours are restricted to areas where the species will be visible to observers on land, and therefore only inshore waters have been mapped.

The photos in this book have been carefully chosen to show key features. Many birds are sexually dimorphic, meaning that males and females have different colouration and patterning. While males often sport brighter, more colourful plumage, it is usually harder to distinguish females of certain species and they can be stumbling blocks for beginners. To help you identify these dimorphic species, this guide gives equal prominence to photos of male and female birds.

Bullfinch (male)

All images show adult birds unless otherwise stated. Where relevant, the photos are also labelled to help with identification:

Male – ♂

Female – ♀

Juvenile – juv.

Immature – imm.

1st/2nd winter – 1st/2nd win.

Breeding plumage – br.

Non-breeding plumage – non-br.

Spring/summer/autumn/winter plumage – spr./sum./aut./win.

Alongside common species, this guide also covers many less-familiar birds of the British Isles. These are species that are either resident or summer visitors with restricted ranges and exacting habitat requirements, or scarce and rare visitors seen during spring and autumn migration.

The photos in this book were taken over many years of birdwatching all around the British Isles. For me, birdwatching started as a childhood hobby and forged a close bond with the natural world. Later, birds became a large part of my career and work too. Birdwatching has been a constant source of comforting natural companionship throughout my life, through every month, season and year.

Whatever your level of interest in birds, they are an integral part of our natural world and key indicators of the state of the wider natural environment. Protecting, monitoring and conserving our avian species has never been more important in the age of the climate crisis. Having an understanding and appreciation of our varied birdlife is a good start, and I hope this book will inspire and encourage others to develop an interest in the birds around us.

Capercaillie *Tetrao urogallus* 74–90cm

Very large, elusive grouse, with broad wings and long, rounded tail. Adult male turkey-like, black, with red eye-wattle and chunky ivory bill. Purple-blue sheen to upperparts, green sheen to breast. Wings dark brown, white shoulder-patch. Belly, flanks and uppertail with variable white feather edging, undertail and underwing white. Tail raised to stiff fan when displaying or in defence. Displays at small communal display ground (lek) or alone, makes cork-popping, hissing noises, can be aggressive. Female intricately marked, fine black-white-brown barring and mottling throughout, whiter below with orange breast-patch. Flight fast, low burst from trees or ground.

♀

Where to see: Scarce, local resident of Scots Pine forest in Scotland, prone to disturbance. Prefers ancient Caledonian forest with mature trees and natural clearings.

♂

Black Grouse *Lyrurus tetrix*   49–58cm

Large, shy grouse. Male stocky, black, like small **Capercaillie**, with blue sheen to neck, red wattle above eye, small black bill and double-hooked long tail. Wings dark brown with white shoulder-spot and wing-bar. Displays at traditional, open 'lek' sites, with black tail fanned, showing outward curling tail feathers and white undertail 'plate'. Flat-backed, long profile in flight, with bold white wing-bar. Female brown, with fine, even bars of cream-black-chestnut and grey throughout. In flight, shows pale wing-bar, notched tail-tip and white undertail.

Where to see: Scarce, declining local resident of Scottish upland moors, pastures, heath and forest edges. Small numbers in N England, Wales. Found on ground or in trees eating buds and shoots.

Ptarmigan *Lagopus muta* 31–35cm

♂ spr.

Beautifully camouflaged, dainty grouse of high mountain, tundra terrain, usually above 1,000m. In winter, both sexes pure snow white, male shows black lores and red wattle above eye. Summer male has white plumage heavily peppered with variable dark grey upperparts. Summer female plumage is an intricate lacing of gold-brown, black-and-white feathers throughout, belly and undertail white. Both show striking white wings and black outer tail wedges in flight. Call a faint, croaking rattle.

Where to see: Scarce, local and declining resident of stony, lichen-covered mountain-top terrain in the Scottish Highlands, particularly in the Cairngorms National Park.

♂ win.

♀ spr.

Red Grouse *Lagopus lagopus*   33–38cm

Large, plump grouse of extensive heather moorland with small head and short feathered legs. Male appears dark at distance. Plumage warm chestnut brown with russet tones and variable small white flecks and chevrons. Bill small, grey; face shows narrow white eye-ring and red wattle above eye. Leg and feet feathering white. Flight strong, straight, rapid beats and glides showing black outer tail wedges and white bars in underwing. Female colder brown, mottled, uneven yellow-buff feather fringes and no eye wattle. Crowing call a distinctive *go-back-go-back-go-back*.

Where to see: Locally common, sedentary, upland resident of NW England, Scotland and in smaller pockets in Ireland and Wales. Conspicuous 'gamebird' species, though often crouches while feeding in favoured open heather moorland, upland bog and rough grazing habitats.

Red-legged Partridge *Alectoris rufa* 34–38cm

Larger, more approachable than **Grey Partridge**, with rounded body, short wings and tail, and bold, distinctive plumage. Sexes similar, pale sandy brown with neat red bill, eye-ring and legs. Cream eyebrow stripe and clean, cream throat bordered by black necklace. Neck and upper breast pale blue-grey, streaked black. Striking flanks of bold blue-grey, cream, black and rufous bars. Plain, down-curved wings in low gliding then rapid wingbeat flight.

Often runs before taking flight. Call noisy, chuffing and clucking.

Where to see: Locally common resident throughout British Isles, densest in E England on sandy, open agricultural and farmland habitat. Introduced as a 'gamebird' in seventeenth century, still reared and released for shooting alongside naturally established populations. Often seen in small family groups (coveys).

Grey Partridge *Perdix perdix* 29–31cm

Small, shy partridge, often found in small inconspicuous groups (coveys). Sexes similar, though male has bolder patterning. Attractive plumage, with an orange-tan face and throat, dainty grey bill, chestnut barred flanks, and pale grey breast with an arched dark chestnut belly-patch, the latter visible only when it stands upright. Back finely barred chestnut and grey. Wings short and barred, tail-sides chestnut-orange, obvious in very low, short flight.

Where to see: Increasingly scarce and localised resident, rare in N and W

♂

regions, and Ireland. Local strongholds in C and SE lowland England. Prefers farmland, open meadows and sheltered field margins with hedgerows.

♀

Quail *Coturnix coturnix* 16–18cm

Tiny rotund gamebird, with streaky warm brown plumage and intricate head pattern of dark brown and pale cream stripes. Back and flanks finely marked throughout, with black-edged white feathering and cream streaks. Male shows variable black central throat-patch; female has pale throat. Elusive nature, rarely seen out in the open, more likely heard singing or seen briefly when flushed. Call often a soft *wrree*, song a very liquid, sharp *wit'wit'wit*.

Where to see: Scarce, widespread summer visitor, Apr–Oct, from Africa.

Patchy distribution with an east coast bias. Frequents open habitats, cereal fields, extensive grasslands and downlands.

Grouse, partridges, quails and pheasants

Pheasant *Phasianus colchicus* 70–90cm

Large, common, introduced gamebird, with long body and very long, stiff tail feathers. Male striking with iridescent green-blue-black head, ear-tufts, extensive red bare skin over face, and white collar. Body plumage golden-brown to coppery-red with intricate feather details including dark flank chevrons and white-centred shoulder feathers edged orange. Tail lighter brown with black barring. Variations in colour and patterning common. Female similar shape but light fawn-brown throughout with dark-centred cream feathers and dark tail barring. Both have small cream bill, legs and feet. Call a loud *kru-kook*.

Where to see: Common and abundant resident. Millions are released each year by shooting fraternity. Found in farmland, woodland, large gardens, even reedbed habitats throughout UK.

Brent Goose *Branta bernicla* 55–62cm

Small **Mallard**-sized goose. Pure black head, neck and breast, only broken by variable, white neck patch. Upperparts and belly grey-brown, pale flank barring, wing-tips black contrasting with pure white rear-end. Juvenile lacks white neck patch. Two forms occur, **Dark-bellied Brent Goose** and **Pale-bellied Brent Goose**. Dark-bellied is commonest, from Russia and Siberia. Pale-bellied arrives from two populations: Svalbard birds overwinter in Northumberland; Greenland and Canadian Arctic birds stop in Iceland then W Britain and N Ireland. Pale-bellied are browner-backed, paler, whitish-grey through underparts. Appears very black and white in flight. usually fly

Pale-bellied

in patchy groups rather than neat lines or Vs. Flock calls low, croaky choruses.

Where to see: Both forms are locally numerous winter visitors, Oct–Mar. Frequent mainly coasts, estuaries, saltmarshes, and also graze wet grasslands and fields.

Dark-bellied

Canada Goose *Branta canadensis*   80–105cm

Large, long-necked goose with striking plumage enabling easy identification. Head and neck black, with oval white face-patch from neck up towards ear like a chin-strap. Upperparts tawny-brown, with well-defined pale feather-edge barring throughout. Underparts pale brown-cream from breast to pure white undertail, with darker, dusky barring to flanks. Bill and legs black. In flight, blackish tail and rump broken by white crescent. Voice a loud, clear honking. Often flies in tight flocks.

Where to see: Common, widespread resident, introduced from North America more than 300 years ago. Found in lowland rural and urban habitats, on ponds, lakes, marshes, estuaries, grasslands, and river valleys.

Barnacle Goose *Branta leucopsis*   58–70cm

Small, short-necked goose with white, grey and black plumage. Face cream-white with black lores, otherwise uniform black through nape, neck and breast. Upperparts heavily barred blue-grey, white and black. Lower breast and flanks contrastingly pale grey-white with fine dark barring towards flanks and rear. Tail, bill and legs black. In flight, black neck, breast and two-toned darker wings contrast with white belly and face. Call high-pitched bark. Form noisy flocks called 'skeins'.

Where to see: Locally numerous, coastal winter visitor, Oct–Apr (20% world

population overwinters in the British Isles). Greenland birds winter in Scottish Western Isles, Islay and Ireland via Iceland. Spitsbergen birds winter in Solway Firth via Norwegian coast. Russian birds winter in W Europe, some reaching E coast. Prefer coastal grasslands, saltmarshes, estuaries.

Greylag Goose *Anser anser* 74–84cm

Large, bulky, familiar species with an ungainly stance. Large head, chunky orange bill, thick, fleshy pink to orange legs. Fairly unmarked goose, though dark grey-brown upperparts show pale whitish-buff bars, also evident through flanks. Breast and belly cleaner, paler grey. Rear-end strikingly white, dark brown central tail-band and pale grey rump evident in flight. Underwing dark with pale grey contrasting forewing. Call loud, donkey-like honking, strong cackling, very vocal in flight. Most domestic geese are descendants of Greylag.

Where to see: Fairly common, widespread resident and winter visitor (includes many feral populations). Entire Icelandic population overwinters in Scotland and NE England, Oct–Mar. Found near freshwater lakes, meadows, farmland, urban wetlands. Winter flocks on estuaries, rivers, lakes, reservoirs and agricultural lands.

Taiga/Tundra Bean Goose *Anser fabalis/serrirostris*   66–88cm

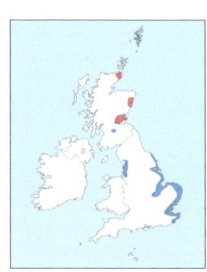

Hard to distinguish from each other and **Pink-footed Goose**. Taiga and Tundra are browner, longer neck and bill, larger head. Taiga has wedge-shaped orange patch over most of upper mandible, contrasting with black, straight-edged lower mandible. Head and upper neck dark brown leading into mid-brown upperparts, pale barring to back and flanks. Breast and belly paler brown-grey, rear-end white, legs orange. Tundra relatively short neck, chunky sloping bill accentuated by longer head, **Whooper Swan**-like profile. Black bill, small oval orange patch near rounded bill tip. Both show dark wings in flight. Quieter than most geese.

Where to see: Both scarce, local winter visitors from N Russia and Siberia, Oct–Apr. Small numbers of Taiga overwinter in Scotland and E coast England. Tundra rarer, E coast. Frequent rough, wet coastal fields, meadows and arable land.

Tundra Bean

Taiga Bean

Pink-footed Goose *Anser brachyrhynchus* 64–76cm

Medium grey-brown goose with small, round, dark brown head, short neck and comparatively short, stout dark bill with variable pink patch towards the tip. Back dark grey-brown with strong white-pale grey feather fringe barring throughout. Underparts contrastingly soft brown-buff with dark brown barring through flanks, darkening heavily towards clean white rear-end. Legs pink, though hard to determine in poor light. In flight, flanks darker and underwing-coverts paler than both **Bean Geese**. Voice a loud, cackling chorus, often includes higher pitched *wink-wink* notes in flight. May form very large flocks.

Where to see: Fairly common, winter visitor from Greenland and Iceland, Sep–Apr. Mainly localised along E coast Scotland, England and Ireland. Prefers large arable fields, coastal mudflats, estuaries, lakes and marshes.

White-fronted Goose *Anser albifrons* 64–78cm

Medium, compact goose with two sub-species: 'European' ssp. *albifrons* and 'Greenland' ssp. *flavirostris*. Solid head structure, pink bill on ssp. *albifrons*, orange with pale tip on ssp. *flavirostris*. Diagnostic bold white forehead blaze. Dark brown-grey back, less grey than **Greylag**, fine pale barring. Clear white stripe along flank. Head and underparts brown-grey, black mottled belly-bars and patches, more extensive belly on ssp. *flavirostris*. Rear-end white, legs and feet rich orange. Juvenile lacks white blaze. In flight, uniform brown above, paler greyish-brown below, belly-patches conspicuous. Call cackling and yodelling. Often in small flocks.

Where to see: Scarce, local winter visitor, Oct–Mar. 'Greenland' birds winter in Ireland, Scottish islands and W Wales; 'European' birds in England. Frequents wet grasslands, coastal marshes, arable fields, lakes and river valleys.

Mute Swan *Cygnus olor* 145–160cm

Very large, familiar, white waterbird, with long, thick neck held straight or in an 'S' shape, head typically angled down. Bill chunky, orange-red, thin black trim, black tip and base. Conspicuous black knob at forehead, very bulbous on males during breeding season, less obvious in females, absent in juveniles. Ungainly on land, graceful on water with smooth, effortless movements, wings often held slightly raised, particularly when threatened or being assertive. Juvenile washed-out, grey-brown throughout with greyish-pink, black-edged bill. Call mainly weak grunts and hisses. Unique rhythmic 'airy wheeze' sound produced in flight from powerful, slow wingbeats.

Where to see: Fairly common, widespread resident and winter visitor from Europe. Frequents lakes, slow rivers, reservoirs, large ponds and sheltered coastal waters. Also open habitats including fields in winter.

juv./imm.

Geese, swans and ducks

Bewick's Swan *Cygnus columbianus* 115–127cm

Small, compact; short neck, rounded head. Bill large, sloping, slightly concave top profile. Diagnostic, variable small yellow rounded patch at base of black bill. Tail square-ended; legs short and black. Juvenile grey-brown, with dusky pink-grey wedge on black bill.

Where to see: Scarce, locally uncommon winter visitor from Siberia, Oct–Mar. Found at traditional sites, mainly S England. Open, wet grassland, inter-tidal and flooded arable sites, open freshwater by dusk. Calls various deep honks, less bugle-toned, higher pitched than Whooper Swan.

Whooper Swan *Cygnus cygnus*   145–160cm

Larger than **Bewick's Swan**; flattened forehead, longer, tapered bill. Neck long, held straight. Bill shows variable, large wedge-shaped yellow patch, small black tip and base trim. Tail square-ended; legs short and black. Juvenile grey-brown, pale pink-grey wedge on black bill. Calls include loud, deep bugles, honks and trumpeting.

Where to see: Scarce, locally numerous winter visitor from Iceland, Oct–Mar, scarce breeder. Much larger and more widespread wintering population than Bewick's. Open freshwater, wet grasslands, flooded arable sites, intertidal zones and mudflats in England and Ireland. In Scotland, breeds in remote upland lochs.

1st win.

Egyptian Goose *Alopochen aegyptiaca* 63–73cm

Larger than **Shelduck**, well built, with straight stance and proportionally long legs. Sexes alike with dark orange-brown upperparts, black, dark green and white wings and black tail. Head and underparts dusky grey to yellow-cream, unmarked. Neck shows small dark chestnut band; head has a prominent chestnut-brown patch surrounding yellow eye. Bill small, pink and black; legs dull red to pink. Beware notable variations in adult plumage, including all white head. In flight, wings very pied black and white from above and below, like Shelduck.

Where to see: Resident, local, introduced from Africa in eighteenth century. Found mainly in eastern counties of S and C England. Frequents urban and rural parks, lakes, reservoirs and marshes; quite disturbance tolerant. Sometimes perches in trees.

Shelduck *Tadorna tadorna* 58–67cm

♀

Like a small, heavy-chested goose with striking plumage. Mainly white with contrasting dark green (almost black) head and neck, broad chestnut-brown breast-band, and black shoulders, wings and central belly stripe. Bill large, bright red-pink; legs bubblegum pink. Adult male has a bulbous red-pink knob on forehead. Very pied in flight with black head and wings contrasting with white plumage. Calls include a smothered whistle and nasal *ga-ga-ga*.

Where to see: Fairly common, widespread resident and winter visitor, though many birds leave for the Baltic states and Germany to moult Jul–Oct. Remaining birds, often family groups, increase with additional birds from N Europe Oct–Mar. Prefers coasts, mudflats, marshes, estuaries and some freshwater wetlands.

♂

Mandarin Duck *Aix galericulata* 41–49cm

An introduced, exotic-looking small duck. Male distinctive with multicoloured plumage. Solid, sweeping white crescent from eye to back of lower neck, long, sweeping orange cheek plumes, violet breast bordered by vertical black-and-white stripes, white belly, soft brown flanks and flamboyant raised burnt-orange sail along wing. Petite pink bill; legs orange. Female grey-headed with blunt crest, white eye spectacle and bill base. Brown upperparts, flanks brown with bold white spots, breast pale grey with fine white dappling, lower belly white. Bill grey-pink with white tip.

Where to see: Scarce, established Asian species, mainly in S England. Prefers lakes, ponds and freshwater bordered with trees. Nests in tree holes.

♀

♂

Garganey *Spatula querquedula* 37–41cm

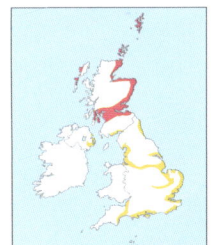

♀

Small, shy dabbling duck, larger than **Teal**, proportionally long, slim bill. Male distinctive, bold white head-stripe, brown finely speckled head and neck, pale grey flanks. Long plume-like black-and-white feathers, rear buff with fine dark spotting. Female similar to female Teal, but colder; body plumage more boldly marked, head pattern stronger with dark eye-stripe, dark cap and whitish spot at base of grey bill. In flight, dark wing-panel bordered by white above and along trailing edge, like **Mallard**. Male forewing pale blue-grey. Both show black leading edge and white armpit to underwing.

Where to see: Scarce, local summer visitor and passage migrant, scattered locations England, Scotland, Mar–Oct. Britain's only summer migrant duck; winters in Africa. Frequents freshwater marshes, wet meadows and pools.

♂

Shoveler *Spatula clypeata* 44–52cm

Short-necked, **Mallard**-sized dabbling duck, with very distinctive large spatula-shaped, sloping bill. Male has dark green head, yellow eyes, black bill, white breast and rich chestnut flank-patch. Long black-and-white feathers along back. Forewing powder blue, obvious in flight, and no white trailing edge. Female similar to Mallard female, but richer mottled brown plumage throughout and large orange shovel-shaped bill. In flight, pale grey forewing, no white trailing edge to dark wing-panel. Overall shape and weighty spatula bill make for straightforward identification. Voice quite clipped, quiet quacks. Very agile in flight.

Where to see: Locally common resident and common, widespread winter visitor from N Europe and Siberia. Frequents coastal and inland waters, large lakes and marshes.

Gadwall *Mareca strepera* 46–56cm

Medium-sized dabbling duck, smaller and slimmer than **Mallard** with squarer head. Male finely speckled grey, buffy-brown head, conspicuous black rear and rump. White wing-patch evident at rest and in flight, chestnut-and-black wing-patches only noticeable in flight. Bill dark grey-black; legs orange. Female like female Mallard, but orange sides to bill, colder-toned plumage, white wing-patch, entirely white belly. Well-marked with streaky pale feathers with dark brown centres. Head plainer, grey with soft dark eyeline. All show yellow-orange legs.

Where to see: Locally common resident, winter visitor from NE Europe and Iceland, Oct–Mar. Highest densities S and C England and Ireland. Often found in pairs or small groups on large wetlands, lakes, reservoirs and freshwater habitats with ample vegetation. Less frequently on brackish and coastal habitats.

Wigeon *Mareca penelope* 45–51cm

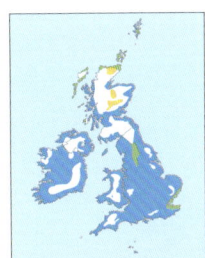

♀

Medium-sized short-legged duck with rounded head and short-bill. Male distinctive with chestnut-red head, creamy-yellow forehead, dusky pink neck and breast, and dove grey body. Rear-end white and black with black pointed tail. Legs grey, bill pale grey with black tip. White upper forewing-patch obvious in flight. Female has pale-fringed scaly brown upperparts, warm brown head, darker around eye, variable orange-brown underparts, flanks, and contrastingly white lower breast and belly. Pale belly obvious when feeding on land. In flight, female has no white patch to wing. Male call an urgent, whistling *whee-ooo*. Flocks can be very vocal.

Where to see: Locally common winter visitor, Sep–Mar, rare breeder N England and Scotland. Frequents coasts, estuaries, wet grasslands, marshes, lakes and reservoirs.

♂

Mallard *Anas platyrhynchos*   50–65cm

♀

Large, familiar duck with comparatively long body and long, deep bill. Male has glossy green head, yellow bill, narrow white collar, mauve-brown breast, greyish body and black-and-white rear. Legs orange. Female mottled buff-brown, with broad, buff-fringed, dark brown feathers throughout and orange-and-brown bill. Note brown belly, not whitish like other female dabbling ducks. Light brown head shows dark brown eyeline and crown. All show violet-blue wing-panel bordered by narrow black-and-white bars, obvious in flight. Calls varied, including familiar *quack*. Highly tolerant of humans; domesticated escapees exist alongside wild ones, the former showing highly variable plumage.

Where to see: Common, widespread resident and winter visitor. Found in varied wetland habitats throughout the British Isles from coasts and estuaries to lakes and inland reservoirs.

♂

Pintail *Anas acuta* 51–66cm

Large, slender, long-necked dabbling duck. Male has distinctive long, pointed black central tail feathers, pale greyish plumage, dark brown head and neck, white breast extending up hindneck in narrow stripe. Rich cream patch near rear, black wing-patch and black undertail. In flight, dark wing-panel bar bordered by white along trailing edge. Bill pale grey and black. Female shorter-tailed, scalloped buff-brown above, paler below with bold dark brown chevrons, whiter towards rear. Head plain brown, bill grey and slim. Both show grey legs. Slender shape, head, pointed tail and long neck give distinctive flight profile.

Where to see: Locally common, widespread winter visitor from N Europe; rare breeder. Mainly found on coastal marshes and estuaries but also inland lakes and reservoirs.

Teal *Anas crecca* 34–38cm

♀

Small, shy dabbling duck. Male distinctive, dove grey above and below, fine black flecks throughout, dark speckling to buff-cream breast. Head chestnut, broad green patch bordered by thin cream lines. Cream-yellow triangle bordered black at rear-end, black-and-white horizontal stripe along wing. Female like tiny female **Mallard** with bolder, mottled, pale-fringed cold-brown feathering throughout. Small white streak to side of tail. Both show broad white bar, and black and bright green panel to wings during fast, vertical, erratic flight. Male calls high-pitched, thin, far-reaching *cree* whistle or relaxed *prip-prip* repeating.

Where to see: Common, widespread winter visitor, Oct–Mar, from Iceland, NW Europe, Siberia, preferring lakes, reservoirs, gravel pits, estuaries and open coastal marsh habitats. Feeds along edges, often hidden in vegetation. Scarce breeder, NW England and Scotland.

♂

Green-winged Teal *Anas carolinensis* 34–38cm

Very like **Teal**, but male has vertical white crescent stripe on breast side and no horizontal white stripe along edge of wing. Large green facial band doesn't have the continuous cream edging seen on Teal. Male has creamy breast with more obvious black spotting. Female hard to tell from female Teal, but often stronger facial markings and white 'loral' bill-base spot. In flight, both sexes show green speculums and warm buffy mid-wing bar which is distinctly white on Teal.

Where to see: Scarce annual migrant from North America, mostly autumn, Sep–Mar, many readily overwinter. Freshwater lakes, pools, also estuaries and marshes.

Ring-necked Duck *Aythya collaris* 37–46cm

Resembles **Tufted Duck**. Profile shows high-peaked hind crown. Male black with large grey white-edged side panel. Bill slate grey with white 'ring' band and contrasting black tip. Bill base edged white. Female like Tufted Duck female, but with distinct profile and bill pattern. Plumage cooler brown, whitish-brown wash to face, throat and underparts.

Face shows whitish eye-ring and stripe behind eye. First-winter has dark bill with one white band. All show pale greyish wing-bar in flight.

Where to see: Rare, annual passage migrant from North America. Mostly found in autumn, Sep–Mar; some overwinter. Freshwater habitats.

♀

Medium-sized diving duck with a high crown, gently sloping forehead and longish broad bill, giving a distinctive profile. Male has chestnut head and neck, red eye, blue-grey band across dark bill and pale grey body, contrasting with black breast and rear. Female dull grey-brown with warmer-toned brown head and breast. Head, neck and breast mid-brown with whitish areas around eye, bill base and throat. Dark eye and lore-patch. In flight, upperwings lack strong contrast, appearing uniform grey; wings short, heavy-bodied. Mainly feeds at night, sleeps by day.

Where to see: Locally common resident and winter visitor from Europe and Russia. Can form large flocks. Prefers freshwater marshes, lakes, well-vegetated reservoirs and gravel pits.

♂

Tufted Duck *Aythya fuligula* 40–47cm

Medium-sized, round-headed diving duck with a long crest or 'tuft' hanging from hindcrown down towards nape. Male very black and white with black crest, head and body contrasting with well-defined white flanks and belly. Black head has violet gloss in good light. Female rich brown throughout, darker on upperparts, paler and mottled along flanks; short brown crest. Can show variable whitish patch at bill base, and on undertail. Both have golden-yellow eyes, blue-grey bill with paler grey band near black tip. In flight, both dark above with obvious long white wing-bars.

Where to see: Most common diving duck, a widespread resident and common winter visitor, Sep–Apr, from Iceland and N Europe. Likes freshwater habitats, lakes, rivers, gravel pits and reservoirs. Also sheltered bays and coasts in winter.

Scaup *Aythya marila* 42–51cm

♀

Larger than **Tufted Duck**, rounder head, no crest. Male well-defined black head, often showing green sheen, black breast and rear. Flank-sides clean white, back also white unlike male Tufted Duck; fine grey barring throughout. Eyes yellow, bill greyish-blue with a very small black tip and no pale band. Poorly marked female and juvenile trickier to tell from female Tufted Duck, but variable, often obvious large white patches around bill base.

Adult female has pale brown-grey tones to mottled brown back and pale flanks. In flight, all have white wing-panels from above.

Where to see: Scarce winter visitor from Iceland and N Europe, Sep–Apr. Favours saltwater habitats, sheltered coastal bays and estuaries throughout the British Isles. Occasionally inland on lakes and reservoirs, in small numbers.

♂

Eider *Somateria mollissima* 60–70cm

♀

Large, heavy duck, thick neck, wedge-shaped head and bill making for distinctive profile. Breeding-plumaged male has distinctive white upperparts, white patch to otherwise black flanks, belly and tail. Head white with black crown and forehead, orange bill with pale grey tip. White nape has pistachio-green patch, white breast shows a rose flush. Stubby, square tail often raised. Eclipse and first-winter males dark brown-black, mottled white. Female rich brown throughout, fine dark barring through back, breast, flanks. Bill grey-green, yellow tipped. Evocative cooing *aa-ooh* from courting males, often in social rafts at sea.

Where to see: Locally common resident, coastal, marine areas from N Scotland and Ireland to N England. Common winter visitor from Scandinavia, throughout the British Isles, including S and E England coast.

♂

Velvet Scoter *Melanitta fusca* 51–58cm

♀

Large, heavy-set sea-duck, angled, sloping forehead profile, rather long, slightly upcurved bill. Dark red legs, thick neck, short tail, often held cocked. Both sexes dark brown to black throughout with large white wing-patches, sometimes evident at rest, more conspicuous in flight, aids identification from **Common Scoter** flocks. Male black with white 'tick' below eye, bright orange to bill. Female browner throughout, some show white cheek-patch, all have diffused white patch between eye and dark bill.

Where to see: Scarce winter visitor and passage migrant from Fennoscandia. Like Common Scoter, found in open, rough seas through to sheltered bays and islets. Mainly offshore along E coast, from N Scotland to East Anglia, in small groups, pairs or associating with larger Common Scoter flocks.

♂

Common Scoter *Melanitta nigra* 44–54cm

♀

Large, elegant sea-duck with barrel chest, pointed tail and slim neck, both often held upright. Male all black with dark brown wing-tips. Bill wedge-shaped, dark with knob to base and narrow yellow patch along top. Female dark brown with two-toned head of dark brown above, pale brown below. Bill dark. All dark wings in flight. Often in large groups, flies in low, long lines offshore.

Where to see: Common, fairly widespread coastal winter visitor and passage migrant, uncommon summer visitor from Iceland, N Europe and Siberia. Very rare breeder, lochs of N Scotland and Ireland. Small numbers can be seen all year, favouring open sea, sheltered large bays, close inshore waters in winter. Largest numbers winter off NE Scotland and Wales.

♂

Long-tailed Duck *Clangula hyemalis* 40–47cm

Small, **Tufted Duck**-sized sea-duck, dumpy-neck, rounded head, short, deep bill. Complex plumages. Winter male striking, clean mix of chocolate brown and white, very long, tapered tail feathers. Head white, mushroom-brown face, large blackish cheek-patch. Bill black, pink band near tip. Winter female mid-brown, white face, neck and underparts. Large brown cheek-patch, grey-blue bill. Summer male and female browner, mottled with less white. In flight, male has distinct profile, dark breast-band on white underparts, long, protruding tail. Flies low over water, often found in winter 'rafts' at sea. Call evocative, yodelling *ar-ar-ard-oow* by displaying males.

Where to see: Scarce, locally uncommon winter visitor from Arctic to mainly N and E coast England and Scotland, Sep–Apr. Found mainly offshore and also in sandy, sheltered bays.

♂ win.

♀ win.

♂ 1st win.

Goldeneye *Bucephala clangula*   42–50cm

Fairly dumpy, active diving duck with large rounded head and peaked crown. Adult male unmistakable, with black-and-white body, dark green-black head and large white lore-spot between golden eye and black bill. Female grey-brown with contrasting chocolate brown head, white wing-patch and collar, pale golden eye and orange-tipped dark bill. Juvenile similar to female but with brown eye, no white collar. In flight, all show white innerwing-panel to upperwing. Male display includes throwing head backwards across back.

Where to see: Fairly common and widespread winter visitor from Scandinavia and NE Europe, Sep–Apr. Rare breeder, in Scotland using specifically designed nest boxes and tree holes. In winter, found in fresh and saltwater habitats from lakes and reservoirs to rivers, estuaries and coasts.

Smew *Mergellus albellus*   38–44cm

Small 'sawbill', only marginally bigger than **Teal**. Slender body, neat head with steep forehead and slightly crested crown. Striking, well-defined plumage in both sexes. Male shows clean black-and-white plumage with pale grey flanks. Head white with small black face-patch and black nape-stripe. Female and juvenile show rufous-brown head, contrasting with extensive white cheeks and throat, pale grey-brown body and white belly. Both have small, compact, serrated grey bills and obvious white innerwing-patches to upperwing in flight.

Where to see: Scarce, local winter visitor from N Scandinavia and Siberia, Nov–Mar, mainly in C and SE England. Found mainly on freshwater inland lakes, large ponds, reedbeds, reservoirs, gravel pits and rivers.

♀

♂

Goosander *Mergus merganser* 58–68cm

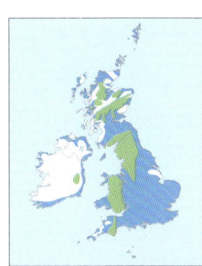

Large **Mallard**-sized diving duck, long body, long, tapering red, hook-tipped bill. Male distinctive with green-black head and upper neck, contrasting white underparts and wing-panel and black back. Female similar to **Red-breasted Merganser** female, but thicker-set, larger head, clear-cut contrast between chestnut-brown head and upper neck, and pure white foreneck and breast. Upperparts and underparts grey-brown. Can show smooth, tapered head profile or steep, bulbous forehead and nape profile. Eyes dark, legs bright red. Flight strong with elongated profile, large white wing-patches.

Where to see: Scarce, local resident, mainly NW. Locally common, widespread winter visitor from Fennoscandia and NE Europe. Prefers inland lakes, reservoirs, lowland rivers in winter, upland slow rivers and lakes near forests in summer. Nests in tree holes.

Red-breasted Merganser *Mergus serrator*   52–58cm

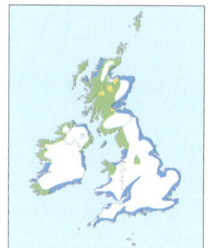

♀

Fairly large, slender-bodied 'sawbill' with bright red bill much thinner, longer, more serrated than in other ducks. Male intricately patterned, green-black head and upper neck, spiky crest, white neck collar, buff-brown breast peppered with dark spots. Back and shoulder black, small cluster of white spots on shoulder. Wings white, flanks and tail pale grey. Female has orange-brown head, crest and upper neck, softly merging into uniform grey-brown lower neck and body. Both have red eyes and legs. In flight rather elegant, elongated, with white innerwing-patches.

Where to see: Uncommon, local resident, mainly NW, Scotland and Scottish islands. Fairly common, widespread winter visitor, Aug–Mar, from Iceland and N Europe. Prefers coastal habitats, estuaries, open sea in winter, occasionally rivers and lakes.

♂

Nightjar *Caprimulgus europaeus* 26–28cm

Cryptic plumage, slender body, long wings and tail, large, flat-topped head, oversized owl-like dark eyes. Camouflaged, bark-like plumage of intricate grey, black and brown-buff markings. Short legs, tiny bill opens to massive gape to catch prey. In flight, male shows white patches near wing-tips (lacking in female and juvenile) and outer tail. Song unusual, reeling *churr* which can last for hours, interspersed with wing-claps. Roosts horizontally on branches or ground by day; agile aerial insect hunter from dusk.

Where to see: Scarce, local summer visitor, scarce passage migrant Apr–Sep, winters in Africa. Breeds mainly S England, East Anglia, also patchily in Scotland, N England and Wales. Prefers dry, open, stony clearings with scattered trees, heaths, woodland edges. More varied, often coastal habitats on migration.

Alpine Swift *Tachymarptis melba* 20–23cm

Large, dark brown and white, scythe-winged swift, with powerful flight. Longer-winged, heavier-bodied and significantly larger than **Swift**. All chocolate brown above, readily identified by large white belly-patch and white throat below, separated by brown necklace. Tail and undertail brown, tail tapered, only subtly forked.

Where to see: Rare annual vagrant to mainly S and E coasts, from S Europe, most Mar–Jul. May roost on churches, tall buildings, otherwise airborne.

Swift *Apus apus* 16–17cm

All dark, scythe-shaped, with long, stiff wings, tapering, forked tail. Adult dark brown-black, small, diffused pale chin patch. Juvenile paler; can appear scaly, with pale feather fringes. Often at height in fast flying 'screaming' parties or hunting insects low over wetlands. Perches only when nesting in old building eaves, roof spaces, nest boxes, even sleeps on the wing.

Where to see: Locally common, widespread summer visitor, Apr–Sep, mostly S and E; winters in Africa. Variety of habitats, particularly suburban and urban villages and towns.

Great Bustard *Otis tarda* 75–105cm

Unique huge, sturdy bird, large, heavy body, long legs, short tail, long, thick neck. Strutting, horizontal profile. Adult male pale, plain grey head, long white whiskers drooping from base of chunky grey bill, bright orange-red neck and breast; drabber in non-breeding plumage. Upperparts intricate mix of black, white, copper bars. Underparts white. Large tail often held flat, raises in display to copper-red and black-barred fan. Female similar but smaller, duller, without copper-red neck or white chin whiskers. In flight, like giant goose, conspicuous thick black trailing edge, white bar across upperwings. Shy nature, often found in groups, slow, measured walking, ground feeding.

Where to see: Formerly widespread in open grasslands and farmland throughout Britain. Extinct here by early nineteenth century. Small reintroduced population in S England.

Cuckoo *Cuculus canorus* 32–34cm

♀

Collared Dove-sized, with hawk-like appearance. Readily perches, with long, rounded tail raised, long wings drooped and body tipped forward on short legs. Adult slate grey above, paler grey head, neck and upper breast. Underparts white with neat dark barring, tail with white spots. Eyes, eye-ring, legs and bill base yellow. Bill short, downcurved, grey. Female may show buffy neck, occasionally entirely red-brown with heavy barring ('hepatic' morph).

Juvenile browner with buff feather fringes. Lays eggs in other birds' nests, such as **Reed Warbler**. Once hatched, Cuckoo chick ejects host's eggs so is reared alone by host parents. Male call a far reaching *cu-koo* repeated. Female call a bubbly *trill*.

Where to see: Uncommon, widespread summer visitor, Apr–Sep; winters in Africa. Mainly reedbeds, also heaths, farmland and moorland.

juv.

♂

Rock Dove/Feral Pigeon *Columba livia* 30–35cm

Feral Pigeon

Smaller than **Woodpigeon**, with true, wild and common 'feral' forms. Wild Rock Dove pale grey-blue above and below, darker grey head, iridescent green/purple neck-patch. Two bold black bars obvious on closed wing. Eyes orange, legs pink, bill short, grey with swollen white cere. In flight shows dark narrow wing-bar, white rump, dark trailing edge to wings, dark tail-tip, white underwing. Town or feral pigeons and domestic 'fancy' pigeons show huge plumage variations from all white to pied, black, grey, brown. Call deep, purring *coo*.

Where to see: Resident wild birds have small populations along coastal cliffs, rock faces, caves in N Scotland, Scottish islands and W Ireland. Interbreeding with feral pigeons confuses true distribution. Feral pigeons common, widespread, towns, cities, parks, farmland and coasts.

Rock Dove

Stock Dove *Columba oenas*   28–32cm

Smaller than **Woodpigeon**. Dainty, small-headed dove, fairly short wings and tail. Plumage cool grey, no white, iridescent green neck-patch, rose-purple flush to upper breast and black, broken double wing-bar. Eyes beady, dark, bill small, pinkish with a yellow-white tip. Legs pink. In flight, black trailing edge notable to upperwing and tail, broken dark bars on inner wing. Underwing pale grey with dark trailing edge. Direct flight often with more flapping than Woodpigeon. High rises and slow, steep, downward glides, wings held in 'V' when displaying. Call gentle, deep rhythmic *ooo-woo* repeated.

Where to see: Common, widespread resident, scarce in N Scotland and W Ireland. Found in range of habitats, from lowlands to uplands, farmland, woodland, mature parkland, cliffs and quarries.

Woodpigeon *Columba palumbus*   40–42cm

Large, portly pigeon, broad wings, long tail, small head. Adult grey-blue upperparts and head, soft pink-mauve breast, pale grey-white belly and undertail. Distinctive white patch and iridescent green area on neck. Bill small, stout, pale pink, yellowy tip and white cere at base. Eye small, straw yellow, giving startled expression. Legs pink and short (walks with waddling gait). In flight, upperwing shows large white 'crescent' and black outer wing wedge. Tail broad black band at tip. Flight fast, direct, with 'rise and stall' and wing-claps. Often clatters into flight. Song soft, rhythmic five-noted 'cooing'.

Where to see: Very common, widespread resident of farmland, woodland, towns and gardens. Eats grains, shoots, seeds, plants (brassicas, ivy). Seen as farmland pest, increasingly confiding in urban habitats. May form huge feeding flocks in arable fields.

Turtle Dove *Streptopelia turtur*   26–28cm

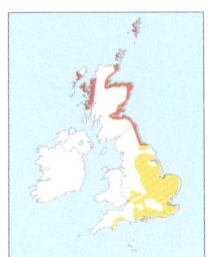

Small, slim, diamond-tailed dove, smaller than **Collared Dove** with unmistakable plumage. Adult has scaly patterned orange-brown edges to dark centred feathers on wings. Proportionally small head, 'dove' grey becoming dirty pink through unmarked neck and breast, whiter towards belly and undertail. Face plain, neat red eye-ring around pale eye. Neck has small black-and-white striped patch. Small bill grey, legs pink. Grey uppertail striking when fanned, with black band and broad white tips. Juvenile browner throughout with no neck-patch. Underwing dark, wings often swept back in flight. Song a mournfully soft, far-reaching purring *curr-currrr-curr*.

Where to see: Scarce summer migrant from Africa, Apr–Sep, mainly restricted to E and S England but occurs along E coast up to Scottish islands during migration.

Collared Dove *Streptopelia decaocto* 31–33cm

Neat, rather plain, pale dove with slender body, long tail and small rounded head. Adult pale beige-buff throughout with faint blue-grey and pinkish hues and conspicuous narrow black collar mark. Wing shows black primaries, evident at rest and in flight. Bill grey, legs pink, eyes blood red, appear dark at distance. Juvenile similar, duller with no black bar on collar. Tail buff-grey with broad white tips either side of buff centre. Lacks black tail-band of **Turtle Dove**. Song a soft, melancholic, repeated three-note *doo-dooo-do*. Glides on fanned wings during display.

Where to see: Common, widespread resident, rapidly expanded range since the 1960s. Favours areas around human habitation, parks, gardens, woodland, towns, villages and farmland.

Water Rail *Rallus aquaticus*   23–28cm

Shy, slender rail, smaller than **Moorhen**, slightly curved red bill, long pinkish legs. Short, pointed tail often held cocked and 'jerked', showing white undertail. Profile can appear rounded when at rest. Adult two-toned, warm brown upperparts and black centred feathers contrasting with slate-grey underparts and black-and-white barred flanks. Juvenile similar, duller brown, dark bill. Call a *kip-kip-kip*, or loud, drawn-out piglet's squeal repeated. Most vocal at night. Creeps lightly around wetland fringes, readily swims, runs for cover when alarmed. Flight weak, restricted, on short wings, neck outstretched, legs trailing.

Where to see: Fairly common, widespread resident, winter visitor from NW Europe. Resident mainly in lowland wetlands, densest population in Ireland. Prefers freshwater margins with dense vegetation, reedbeds, marshes, ditches, also brackish and saltmarsh fringes.

juv.

Corncrake *Crex crex* 27–30cm

Secretive, land-based bird, smaller than **Moorhen**, profile like oversized partridge. Short, rounded rufous wings, black-centred sandy-brown feather 'streaks' to upperparts, pale grey face and underparts. White-edged rusty bars along flanks. Legs long, pale straw-pink; bill short, pale pink. Heard more than seen, song unique dry *crex-crex*, like finger along a comb.

Where to see: Rare summer visitor from Africa, Apr–Sep. Coasts and island localities in NW Ireland, N and W Scotland and Western Isles. Hay meadows, grassland, dry, thick vegetation and iris beds.

Spotted Crake *Porzana porzana* 22–24cm

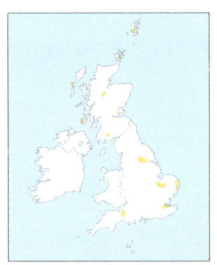

Resembles **Water Rail**, but smaller, rounder, with longer wings, greenish legs, short, straight bill. Dark brown and chestnut above, finely spotted white. Head and throat dark grey, brown cheek, stout yellow bill, reddish at base. Grey-brown below, extensive fine white spots and bars. Pale buff-white undertail. Juvenile duller head and bill. Call a rapid *whit-whit-whit* repeated at dusk and night.

Where to see: Scarce spring May–Jun, and autumn Aug–Oct migrant; rare summer visitor. Rare breeder, Scotland and England. Most in autumn along E coast. Wetland margins, marshes, pools with shallow water and dense vegetation.

Moorhen *Gallinula chloropus*   32–35cm

Medium-sized, round-bodied waterbird with plain dark brown-black plumage. Striking bright red bill 'shield', red bill and contrasting yellow tip. Small red eyes, long green-yellow legs. White, broken bar along flanks and white undertail obvious when tail held characteristically upright. Often seen swimming with bobbing head motion, replaced by jerking tail motion when on land. Juvenile dull, washed-out brown with whitish underparts, dull bill. Calls varied, from soft notes to abrupt *kurr-uk* and croaky *kreks*.

Where to see: Common, widespread resident and winter visitor from N and E Europe. Found in variety of well-vegetated freshwater habitats, lakes, rivers, large and small ponds, marshes, reservoirs and urban parks. Feeds on land and water.

Coot *Fulica atra* 36–38cm

A round-bodied, plump waterbird, slightly larger than **Moorhen**. Distinctive, with plain, sooty black plumage contrasting with a white bill, white forehead 'shield' and red eye. Legs long, robust and yellow-grey, toes long with extended flat lobes. Juvenile brown-grey with paler whitish underparts, lacks white bill and shield. Strong swimmer, often while nodding head. Rather clumsy, heavy in flight. Call often a loud, sharp single *kowk*. Very territorial in summer.

Where to see: Common, widespread resident and winter visitor from N and E Europe. Found mainly on broad, freshwater habitats, large lakes, reservoirs, rivers, gravel pits and marshes.

Crane *Grus grus* 96–119cm

Very large, elegant waterbird with sloping, oval body, very long neck and legs. Body plumage pale grey, black-and-white banded neck and head, red patch to crown. Rear-end a ruffle of extended feathers over wing and tail. Bill dagger-like, bone-yellow, eye variable, red to yellow, legs dark. Stiff-winged in flight, neck and legs fully extended, broad wings with black fingered primary tips and trailing edge, contrasting with grey body and forewing. Juvenile grey-headed, duller. Call a deep, bugling *kroo* repeated.

Where to see: Rare, local resident in England, Scotland, Wales, including some reintroductions. Strongholds in East Anglia, SW England. Scarce passage migrant from Scandinavia and NE Europe, often along E coast, in pairs or small flocks. Prefers open wet grasslands, farmland and coastal marshes.

Little Grebe *Tachybaptus ruficollis*   25–29cm

Small grebe with a short neck, tiny bill and rather dumpy, blunt-ended body. Breeding-plumaged adult has rich chestnut cheek and foreneck, brownish-black cap, neck and back, and paler brown-and-white fluffed-up 'powder-puff' appearance at rear. Distinctive fleshy yellow gape at bill base apparent even at a distance. Non-breeding adult shows very light buff-brown on flanks, breast, foreneck and cheeks, darker on back and crown. Juvenile has black-tipped yellow bill, and black-and-white streaks on cheeks and neck-sides. Rather elusive, shy nature, often heard with a loud, trilling 'whinny' from marginal vegetation before being seen.

Where to see: Locally common and widespread resident. Frequents lakes, ponds, reservoirs and slow rivers; sometimes sheltered coasts in winter.

br.

non-br.

Red-necked Grebe *Podiceps grisegena* 40–46cm

Fairly large grebe, chunkier though smaller than **Great Crested Grebe**, with stockier neck, stouter bill and dark eye. Very striking in summer, with a rich red-brown neck and breast, large dusky-white cheek, neat black crown enveloping dark eye, and dark bill with yellow base. Body dark brown.

Non-breeding adult duller, with diffused smoky-brown through upper cheek and neck, dark brown crown, yellow bill and dusky-white to lower cheek, throat and breast.

Where to see: Very scarce winter visitor from W Europe, Oct–Mar. Found in late autumn and winter on open sea, large estuaries and sheltered bays. Most found off E coast Scotland and England but also S coast. Occasionally found in summer along E coast.

br.

non-br.

Great Crested Grebe *Podiceps cristatus*   46–51cm

br.

non-br.

Largest grebe. Long, slender body and dainty, upright neck giving graceful appearance. Breeding adult striking, with chestnut cheek-ruff and black crest, held down when at ease but dramatically fanned and shaken during display. Upperparts dark brown, underparts and eye area white. Bill slim, pointed, dark pink. Non-breeding adult shows dark brown crown, hindneck and upperparts, clean white face, neck and upper breast. Underparts pale taupe-brown, bill pale pink. Juvenile has black-and-white streaking to face. Courtship involves synchronised water dancing with breast-to-breast contact, growling and barking noises.

Where to see: Common, widespread resident and winter migrant from N and W Europe. Found on large inland lakes and slow rivers with good marginal vegetation. Also on sea, reservoirs and lakes during winter.

Slavonian Grebe *Podiceps auritus*   31–38cm

non-br.

Small grebe with rather round forehead, flat crown and straight, pale-tipped black bill. Striking breeding plumage with black head, bright golden-yellow ear-tufts forming a band each side of head from bright red eye to back of nape. Back grey-black, russet-red neck and flanks. Non-breeding adult black and white at distance, dark cap stopping at eye level, giving clean-lined contrast with clean white lower face and cheek-patch. White foreneck and breast contrast with mid-brown hindneck and back. Flanks mottled dusky grey-brown.

Where to see: Rare breeding bird in Scotland; scarce, widespread coastal winter visitor from E Europe and Iceland. Breeds on northern isolated inland lakes. The most coastal of all the grebes, found in sheltered bays, lochs and estuaries in winter.

br.

Black-necked Grebe *Podiceps nigricollis*   28–34cm

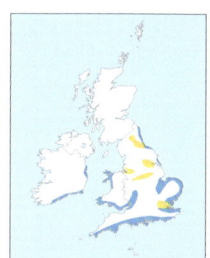

non-br.

Small, dainty grebe with a steeply peaked forehead, slightly upturned black bill and often shows a powder-puff rear. Unmistakable in summer, with jet black upperparts, head, neck and breast contrasting with copper-brown underparts and striking golden-yellow ear-tufts streaming from bright ruby-red eye. Non-breeding adult rather black and white, with dusky white to throat, cheek 'crescent', breast and mottled flanks.

Differs from winter **Slavonian Grebe** at distance by dark crown extending below eye, diffusing white cheek-patch.

Where to see: Rare breeding bird and very scarce winter visitor from N and E Europe. Breeds on freshwater lakes at a few sites in Scotland and England. Winters along mainly E and S coastal sheltered estuaries, occasionally large inland lakes and reservoirs.

br.

Stone-curlew *Burhinus oedicnemus* 40–44cm

Robust wader with large head and long wings. Stout yellow bill with black tip, large bulbous 'staring' eyes, long yellow legs. Pale sandy brown with dark streaking to upperparts, cleaner white below. Bold white stripe above and below eye. White, dark-edged wing-bar, obvious when standing. Large black patches with small white spots to upperwing in flight. Secretive, stands still for long periods. Mostly active from dusk. Call loud, trilling *krrrr-eee*, pipes and whistles.

Where to see: Rare summer migrant from S Europe and N Africa, Mar–Sep. Rare breeder in East Anglia, Wiltshire. Open, stony ground, heaths, short grassland and farmland.

Black-winged Stilt *Himantopus himantopus* 35–40cm

Graceful, distinctive black and white wader with long, thin legs, slim body and fine, needle-like bill. Male black above, variable black to nape and crown, and pure white underparts, undertail and rump. Bill black, legs red-pink. Female brown backed, and reduced, if any black to nape and crown. Juvenile/first winter browner, scaly above with black wings, dusky nape and crown. Legs ochre-yellow. Flight fast, rapid, legs trailing behind body.

Where to see: Rare migrant from S Europe, mainly S and E England, May–Jun. Very rare breeder. Prefers shallow coastal wetland habitats.

Oystercatcher *Haematopus ostralegus* 40–45cm

Unmistakable large, pied wader with black upperparts, white underparts, thick pink legs, and long, straight, carrot-like orange bill. Eyes bright red with orange eye-ring. Obvious white band across black upperwings and 'V'-shaped white rump in flight. White chin-strap evident outside breeding season. Juvenile dirtier, browner-black above, dark-tipped bill and greyish legs. Call piercingly loud *kleeep* repeated and shorter *kip-kip-kip* notes.

Where to see: Common, widespread resident and winter visitor from N Europe. Favours coastal habitats and inland sites in more northerly locations. Prefers upland grasslands, marshes, small islands, shingle beaches, rocky shores. Winter visitors often gather in large flocks on estuaries with resident birds. Many juveniles migrate to Europe in winter.

Avocet *Recurvirostra avosetta*   42–45cm

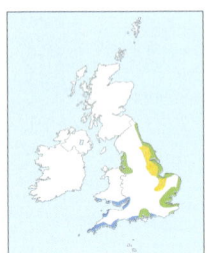

Striking, delicate and unique wader with long grey legs, fine, upcurved black bill and pied plumage of pure white and black. Black cap reaches below eye and along nape, larger black patches to wings, obvious in flight. Juvenile duller, brown and off-white. Feeds by sweeping bill from side to side while wading in shallow waters.

Where to see: Scarce summer visitor, mainly to E and NW England, from S Europe and Africa. Also locally common winter visitor from Europe to mostly SW England. Increasing populations staying all year round with birds from E coast joining migrants in SW in winter. Prefers coastal, brackish lagoons in summer, estuaries in winter.

Lapwing *Vanellus vanellus* 28–31cm

Large, dark green-black-and-white plover, with long, wispy crest plumes. Upperparts dark, green-purple oily gloss in good light, underparts white, undertail chestnut-orange. Male has black forehead, crown, cheek-stripe, throat and bold black breast-band. Female and juvenile similar but shorter crest, whiter on face, throat, neck. In winter, all show pale feather fringes or 'scalloping' to dark upperparts. Legs dark pink, grey in juvenile, bill short, dark. In flight, wings broad, rounded, underwing black and white. Tumbling, rolling display flight in spring. Calls include *pee-wit* and during display a nasal, rising *pwee-y-eeit*.

Where to see: Locally common, wide-spread resident, winter visitor from Russia and NE Europe. Frequents arable fields, grasslands, wet meadows, moorland and salt-marshes. Big flocks gather on estuaries and larger fields in autumn and winter.

Golden Plover *Pluvialis apricaria* 26–29cm

Rotund, heavy-chested plover, fairly short dark legs, dainty round head, small dark bill, black beady eyes. On ground, walks a few steps in upright stance, pauses, pecks after food. Breeding adult fine golden-yellow, brown-and-white speckling throughout upperparts. White tramline stripe on outside of black face, foreneck, breast, belly. Female similar but browner, less contrast. In winter, all ages densely speckled yellow-and-brown plumage throughout, clean white belly and brown speckles, streaks to paler upper breast and flanks. White armpit, dark rump evident in flight. Call a loud, mournful *puu-we*.

Where to see: Locally common resident, widespread winter visitor from Iceland and NE Europe, Sep–Apr. Found on northern upland bogs, moors and grassland (summer); large flocks can gather on arable fields and lowland coastal sites (winter).

juv.

♀ br.

Waders

Grey Plover *Pluvialis squatarola* 27–30cm

Like **Golden Plover** but stockier, with bigger head, heavier bill and more hunched profile. In breeding plumage shows white tramline stripe encompassing black face, neck and upper breast, ending with large white patch. Upperparts spangled grey-silver and black. Juvenile shows browner, contrasting spangled plumage to upperparts. In winter, all ages show pale grey-buff spangling to upperparts (all grey at distance), cleaner, whiter underparts and fine grey speckles to breast. Distinctive black armpit and white rump in flight. Call melancholic, far-reaching *plu-uoo-ee*.

Where to see: Locally common, widespread winter visitor from Siberia and NE Europe, Aug–Apr. Frequents coastal estuaries, saltmarshes and mudflats, often alone or in mixed flocks.

♂ br.

♀ br.

juv.

Ringed Plover *Charadrius hiaticula*   18–20cm

juv.

Rotund plover, slightly larger than **Little Ringed Plover**. Breeding adult, greyish-brown above, white below. Head black and white, buff-brown crown, orange, black-tipped bill. Black bands at forehead and bill base, join as black 'bridle' encompassing dark eye and cheek. Bold black breast-band, short orange legs. Non-breeding adult buff-brown head and breast markings instead of black. Juvenile one brown band through eye. White wing-bar in flight. Call whistly, rising *too-ip*.

Butterfly display flight includes rapid *too-loo too-lou* notes.

Where to see: Fairly common, widespread resident, passage migrant, winter visitor, Mar–Oct. Wintering birds from Europe. Spring passage migrants return to Greenland and Russia; autumn birds to Europe and West Africa. Prefers coastal sand and shingle beaches and estuaries, but increasingly inland on urban gravel pits and inland lake fringes.

♂

Little Ringed Plover *Charadrius dubius* 14–15cm

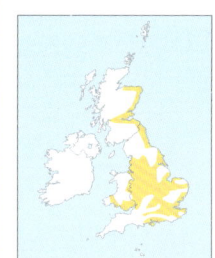

juv.

Small, delicate plover, less rounded than **Ringed Plover** with tapering profile. Plumage similar, but breeding adult shows bright yellow eye-ring and white band above black forehead band and 'bridle'. Legs also pinkish not orange, bill black not orange and black. Non-breeding adult retains reduced yellow eye-ring, black head 'bridle', breast-band becomes brown-buff. Juvenile plainer brown-headed, lacking white band above eye, upperparts with pale 'scalloped' fringes. All ages lack white wing-bar in flight. Call a short, stark *tew*, *pip-pip-pip* and repetitive *gree-a-gree-a* during butterfly display flight.

Where to see: Scarce, local summer visitor from Africa, Mar–Oct, found mainly through England, less common in Wales and E coast Scotland. Prefers inland artificial habitats, gravel pits, quarries, waste ground and reservoirs, also coastal wetlands during migration.

Dotterel *Charadrius morinellus* 20–24cm

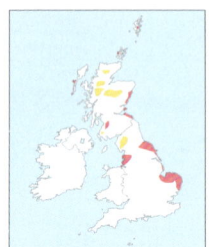

♀ br.

1st win.

Striking, rotund mountain plover. Breeding plumage bold, dark head and crown, white throat, long white stripe above eye, forming 'V' shape from behind. Upperparts plain grey-brown, chestnut breast bordered by white band above and black belly below. Undertail white, legs pale yellow-green. Female brighter than male. Non-breeding adult duller, whiter underparts. Juvenile buff-brown, plainer below, upperparts appear scaly with cream feather fringes. Crown dark, bold creamy stripe above eye, hint of breast-band on peachy-buff breast. Call low purring and abrupt *pwik* notes repeated.

Where to see: Rare, local summer visitor, scarce migrant from North Africa. Mountainous, barren plateau regions in N England and Scotland in summer. More widespread on passage, frequenting coasts, hills and farmland. Usually seen in small groups in spring, often lone juveniles in autumn.

Whimbrel *Numenius phaeopus* 40–42cm

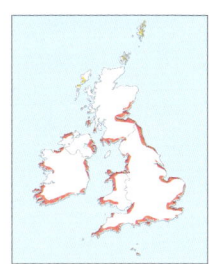

Large wader, smaller and darker than **Curlew**. Rather uniform, densely mottled brown and cream-buff plumage throughout, darker on back, chest and flanks with dark brown feathers edged cream. Belly white, breast and neck with many dark brown fine streaks. Long, slim, dark bill, straighter than Curlew, clearly decurved near tip. Characteristic dark brown eye-stripe and lateral crown-stripe, with contrasting pale central crown-stripe. Legs longish, dark grey. White rump wedge in flight. Call a rapid *tu-tu-tu-tu-tu-tu* trill.

Where to see: Rare summer visitor, scarce, widespread passage migrant. Breeds mainly on N Scottish islands, on coastal, blanket bog and heather moorland. Migrants mainly coastal, heading N in spring and to West Africa and South Africa in autumn.

Curlew *Numenius arquata* 50–60cm

Largest European wader, similar to **Whimbrel** but stockier, longer-legged, with almost comically long, strongly decurved bill. Plumage uniform, strongly mottled brown, grey and buff-cream throughout with plainer whitish underbelly and undertail. Breast, neck and head with fine, dense, dark streaking, head lacking obvious patterning of Whimbrel. White pointed rump wedge in flight. Legs grey. Call distinctive *cur-loo* or *whaups*, song a bubbling trill.

Where to see: Fairly common, widespread resident and winter visitor from Fennoscandia, Sep–Mar. Breeds on upland moorland habitat and coastal marshes, farmland and grassland sites in lowlands. Coastal in winter, mainly on estuaries.

Bar-tailed Godwit *Limosa lapponica* 37–39cm

Like **Black-tailed Godwit** but shorter-legged with slightly upcurved, pink-based bill. Breeding adult male with more extensive brick red plumage through underparts, dark eye-stripe and crown, pale supercilium. Adult female lighter orange-buff with longer bill. Non-breeding adult and juvenile grey-brown, streaked above, whitish below, juvenile with peachy-brown wash. No wing-bars in flight, white rump extends in a wedge onto back, like **Curlew**. Tail has fine dark bars, not solid black like Black-tailed Godwit.

Where to see: Locally common winter visitor and passage migrant from Fennoscandia and Siberia, Aug–Apr. Found around coasts of Britain and Ireland in winter, on estuaries, mudflats and sheltered bays.

♀ br.

♂ br.

non-br.

Black-tailed Godwit *Limosa limosa* 40–44cm

non-br.

Slender, tall, long-necked, long-legged wader. Very long, almost straight bill on small head, giving elegant look. Breeding adult has black-tipped pink-orange bill, rusty-red to apricot head, neck, upper breast, pale supercilium, dark lores and crown. Upperparts mottled grey, apricot, black feathering. Belly white, dark barring, extending along flanks. In flight, white rump, black tail, white wing-bars. Non-breeding adult plain grey-brown above, whitish below. Juvenile like paler breeding adult, peachy neck and breast, cinnamon-brown mottled upperparts.

Where to see: Rare summer visitor, locally common passage migrant, winter visitor from N and C Europe (ssp. *limosa*), Iceland (ssp. *islandica*). Icelandic birds winter in UK, some breed on Scottish islands. European breeds mainly in East Anglia, winters in Europe. Winter birds found on coastal, estuaries and mudflats, mainly SE and SW.

br.

Waders

Turnstone *Arenaria interpres* 22–24cm

non-br.

Medium-sized dumpy wader with short orange legs, broad-based dark bill, and rather pied appearance. Breeding adult shows chestnut-orange and black mottled upperparts. Underparts clean white, head white with smart black face patterning and breast-band. Non-breeding adult more uniform, mottled mid to dark brown feathering through head and upperparts, broad dark brown breast-band, and clean white underparts. In flight, black band across white tail, white wing-bars and three white stripes to central back and shoulders. Call, low, chattering *kuk-a-kuk-kuk* and clipped *tiuw* in flight.

Where to see: Locally common, widespread winter visitor and passage migrant from Arctic Greenland, Canada and Russia. Favours rocky, seaweed-covered coastal habitats, sandy and muddy beaches throughout Britain, often in small groups.

br.

Knot *Calidris canutus*   23–26cm

Medium-sized rather dumpy wader with short, thick, dark bill and short grey-green legs. Breeding adult with brick red underparts contrasting with a mix of grey, black, brown and orange mottled feathering to upperparts. Non-breeding plain pale grey above and white below with light barring to flanks. Juvenile grey above with pale buff-fringed feathering, underparts peachy-buff with fine speckling.

Where to see: Locally common, widespread, coastal winter visitor and passage migrant from Arctic Greenland, Canada and Siberia, Aug–May, largest numbers found Dec–Mar. Can form large, tight feeding and roosting flocks along preferred coastal, estuary and mudflat habitats in winter. Those that pass through on migration winter in W Europe or Africa.

non-br.

Ruff *Calidris pugnax* 26–30cm

♀ juv.

Large wader with small head and bill, elongated body and neck and long legs. Female much smaller than male. In breeding plumage, male has unique, colourful loose-feathered ruff and ear-tufts of black, chestnut and white variations.

♀

Non-breeding male and female often orange-legged, not unlike **Redshank**, but some have ochre or greenish legs. Underparts plain whitish-buff, upper-parts scaly with pale buff-fringed brown feathers. Face plain, with fine dark flecking through crown and variable white areas. Juvenile apricot buff, warmer than grey-toned adult.

Where to see: Rare summer visitor, locally uncommon winter visitor and passage migrant (mainly of juveniles in autumn). Most breed in Fennoscandia and Russia, and winter in Africa. Found on coastal freshwater pools and lakes fringes, brackish lagoons, marshes and fields.

♂ non-br.

♂ br.

Curlew Sandpiper *Calidris ferruginea* 18–19cm

br.

Confusion with **Dunlin** possible but slightly larger and more elegant, with longer legs and longer downcurved bill. Breeding adult shows mottled grey-brown upperparts and rusty to brick red underparts flecked white, not unlike small summer **Knot**. Non-breeding adult and juvenile plainer, dark grey above and white below with strong whitish supercilium and scaly pale-fringed dark feathers to upperparts. Juvenile often has warm, peachy-buff flush to face and breast. All ages show white rump and narrow white wing-bar in flight. Call a thin, high *chirrup*.

Where to see: Scarce, widespread, coastal passage migrant, most in autumn, Aug–Oct. Breeds in Arctic Siberia, winters in Africa. Frequents coastal pools, estuaries, rocky shores, beaches and saltmarshes, often alongside Dunlin.

1st win.

Temminck's Stint *Calidris temminckii* 13–15cm

1st win.

Size of **Little Stint**, but with longer body, tail, wings and a horizontal profile. Legs short, yellowish, bill dark and slightly downcurved. Breeding adult has mottled brown-grey upperparts with variable dark feathers, edged chestnut. Head and upper breast mottled brown-grey, contrasting against white lower breast and belly, like a tiny **Common Sandpiper**. Non-breeding adult plainer, duller brown above with clear brown breast-band. Juvenile grey-toned plain head, weak grey breast- band and darker grey, buff-edged scaly upperparts. Call, fast, chattering *tirr-tirr-tirr*.

Where to see: Scarce passage migrant from Fennoscandia, Siberia and Russia, wintering in Mediterranean and central Africa. Most occur in May on freshwater pools, lakes, reservoirs, and marshes, less often brackish fringe habitats, C and E England.

non-br.

Little Stint *Calidris minuta* 12–14cm

Our smallest wader, with short black bill and legs. Breeding adult (uncommon in Britain) shows bold, blackish feathers with rich chestnut and white fringes to upperparts and clean white underparts. Non-breeding adult much colder, plainer grey-buff above and white below. Juvenile like breeding adult, varied chestnut, black and grey above, with striking white 'tramlines' along back forming a 'V' and white forked stripe above eye. Call quick, sharp *tit*, often repeated.

Where to see: Scarce passage migrant from Arctic Scandinavia and Siberia, Jul–Oct, winters in Africa. Majority are juveniles in autumn, Aug–Sep, along E and W coast. Found on coastal pool fringes, estuaries, brackish lagoons, beaches and mudflats.

Dunlin *Calidris alpina* 16–20cm

non-br.

Small, short-necked, compact wader, long black bill, slightly decurved at tip, short black legs. Breeding adult warm brown upperparts, pale buff and chestnut-edged black feathering. Underparts white, dense, dark streaking through neck and breast, large black belly-patch. Non-breeding adult plain grey above, white below (see **Curlew Sandpiper**, **Sanderling**, **Little Stint**). Juvenile buff-brown, cream-edged feathering above, variable dark streaks and spots to white underparts. White wing-bar, white sides to dark rump in flight. Call a thin, harsh *cree*.

Where to see: Fairly common breeder N and W, common, widespread passage migrant. Birds from Arctic Scandinavia and Russia overwinter. Others breed in Greenland and Iceland and migrate through, to winter in West Africa. Estuaries, beaches and saltmarshes around coastal Britain.

br.

Purple Sandpiper *Calidris maritima* 19–22cm

Stocky, dark, with gently down-curved bill. Non-breeding adult storm-grey above, with pale scaly fringes. White below, grey spots to breast, flanks. Bill-base and short legs orange. Breeding adult mottled brown-chestnut, pale edged feathers above, streaky brown crown, brown streaks and spots to white breast, flanks. Bill and legs dull brown. Juvenile has brighter legs and bill, buffy head, pale scaly upperparts. Call a thin *quit*.

Where to see: Uncommon winter visitor from Iceland, Norway, Sep–Apr. Coastal, localised, mainly NE England and Scotland. Rocky shores, beaches, manmade piers and slipways; often in company of **Turnstone**.

Pectoral Sandpiper *Calidris melanotos* 19–23cm

Like large, robust **Dunlin** or female **Ruff**, fairly long necked with small head. Adult has pale-edged brown feathers above and dense, fine dark streaking through head and breast. Clear contrast between streaked breast band (pectoral line) and white belly, aids identification even at distance. Legs green yellow. Bill stout, dark, paler at base. Juvenile similar but buffier-toned, with chestnut-edges along white-tipped brown feathers through upperparts and pale 'V' on back. Call a low trill.

Where to see: Very scarce annual migrant from North America and E Siberia, most Aug–Oct, anywhere with freshwater wetlands, muddy marshes, small pools, lakes and flooded grassy fields.

Sanderling *Calidris alba* 20–21cm

Close to **Dunlin** in size, but shorter, straighter black bill. Notably larger than **Little Stint**. Breeding adult rufous, brown, cream mottled feathering to upperparts, rusty tones through head, neck, breast. Underparts clean white, legs short, black. Non-breeding plainer, pale grey above, pure white below. Juvenile shows dark brown and grey spangled upperparts, no red tones, though buff through head and collar, white underparts. All dark rump, broad white wing-bar in flight. At close range, note lack of hindtoe. Call a sharp *kuitt*.

Where to see: Locally common, widespread, coastal migrant and winter visitor, Jul–May, peak in May. Breeds in high Arctic, Siberia, Greenland, many winter in Africa. Prefers long, sandy or smooth, muddy beaches. Often seen hunched, running the tideline.

br.

1st win.

non-br.

Woodcock *Scolopax rusticola*   33–35cm

Pigeon-sized gamebird, not unlike oversized **Snipe**, with very short legs and neck, and rotund, chest-heavy body. Cryptic black, rusty brown and buff plumage densely patterned throughout. Bill long, straight, pinky-grey and down pointing. Upperparts 'mottled' brown and rusty brown with pale buff tips. Underparts buff-cream with fine brown barring. Head with distinctly peaked forehead, black bands across crown, a dark lore-stripe and back-set eyes. Often flushes into flight at close range, showing chestnut rump and broad, round-tipped wings. Call a sharp, whistling *tsiwick*. Croaks and grunts during 'roding' display.

Where to see: Common, widespread resident and winter visitor from NE Europe. Mainly nocturnal, secretive, so not often seen. Found in a variety of woodlands and adjacent fields and ditches.

Jack Snipe *Lymnocryptes minimus*   17–19cm

Small, compact, secretive wader, two-thirds the size of **Snipe**, with short, deep-based bill. Legs short, yellowish. Strongly patterned through-out, chestnut, dark brown, buff and cream. Dark brown above, with two pale golden-buff 'tramlines' along back. Almost black eye-stripe, cheek stripe and crown contrast with pale cream supercilium, 'split' with a short dark 'eyebrow' stripe. Often rhythmically 'body bobbing' when feeding.

Where to see: Scarce, widespread passage migrant and winter visitor from Fennoscandia, Russia, Oct–Mar. Prefers wet meadows, reedbeds, fields, ditches, muddy pool edges and wetland fringe habitats.

Snipe *Gallinago gallinago*   25–27cm

Compact, medium-sized, very long straight bill, short yellow-grey legs. Bold buff stripes and chestnut edged brown feathers above. Pale buff-white below, extensive brown flank bars and mottling. Head dark brown eye-stripe and crown, buff central crown stripe. In flight, white trailing edge to wings, rufous tail. Rasping 'sneeze' call. Rhythmic, loud *chip-er, chip-er* on territory. Loud, vibrating tail noises during display 'drumming'.

Where to see: Common, fairly widespread resident, winter visitor from Iceland, Faeroes, N, C Europe. Uplands, lowland moorland, wet grasslands and marshes; most wetland fringes in winter.

Red-necked Phalarope *Phalaropus lobatus* 17–19cm

♀ br.

Slender, dainty little wader, needle-like bill, long neck, mostly observed on water. Breeding female has dark grey upper-parts, bold orange-buff back stripes, rufous neck, white chin, white dash above eye. Underparts white, soft grey upper breast-band, mottled grey flanks. Male similar but duller, orange neck reduced. Non-breeding adult pale grey above, white feather fringes, white below. White head with black cap, slightly down-curving black eye-band. Juvenile like breeding adult but with black-and-white head pattern. Often spins when surface feeding. Call a short *kwit*.

Where to see: Rare summer visitor May–Aug, rare passage migrant Apr–Oct; winters in Pacific Ocean off South America. Breeds in small numbers on marshy pools, mostly Shetland Islands. Visits inland wetland pools and coastal pools in E Britain on passage.

♂ br.

Grey Phalarope *Phalaropus fulicarius* 20–22cm

♂ br.

Similar to **Red-necked Phalarope**, can be equally confiding. Non-breeding pale grey, plainer above, white below with yellow base to black bill. Face white with black hindcrown and straighter black eye-band. Breeding adult rarely seen in Britain, black head with large white face-patch, black-tipped orange bill, cream-and-black striped upperparts and extensive rufous-red from upper neck through all underparts. Juvenile shows patchy black and buff feathering to grey upperparts, buff to neck, slightly raised rear-end to black eye-band. Call a sharp *pit*.

Where to see: Rare, regular autumn migrant from Arctic regions, Sep–Nov, winters in Atlantic off South and West Africa. Most common in autumn along E coast, in sheltered bays close inshore after strong winds.

1st win.

Common Sandpiper *Actitis hypoleucos* 19–21cm

non-br.

Medium-sized wader, readily identified by shape and behaviour. Relatively short neck and legs, slender-body, longish tail. Posture often horizontal. Breeding adult pale grey-brown upperparts, fine dark barring. Underparts white, extending in 'hook' around carpal joint between upperparts and broad, grey breast band. Legs and relatively short, straight bill dull green-yellow. Diffused white supercilium and eye-ring, and dark eye-stripe. Non-breeding plainer above, juvenile more barred above. White wing-bar in flight, no white rump (unlike **Green Sandpiper**). Flight low, with stiff, shallow wing fluttering. Flight call high, loud *swee-wee-wee-wee*.

Where to see: Locally common summer visitor, passage migrant, Apr–Oct. Mainly breeds in N and W upland on lakes, pools, streams and rivers. Freshwater and coastal habitats used on migration. Winters in Africa.

br.

Green Sandpiper *Tringa ochropus* 21–24cm

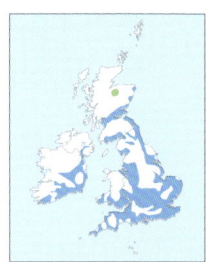

Like **Common Sandpiper** but larger, more contrasting darker upperparts, white underparts. Upperparts and head dark grey-brown, fine white speckling and spots throughout. Dark, diffused, speckled breast. Legs and bill grey green. White supercilium stops at eye, dark lores. Dark wings and white rump in flight, almost **House Martin** like. Call repeated, loud; *kluw-ee, wee-wee*.

Where to see: Fairly scarce passage migrant, rare breeder, uncommon winter visitor, from N and E Europe, Aug–Apr, most winter in Africa. Freshwater and coastal fringe habitats, pools and manmade waterbodies.

Wood Sandpiper *Tringa glareola* 19–21cm

Like **Green Sandpiper** but finer featured, comparatively longer yellow legs, browner-grey upperparts 'spangled' with pale buff spots. Underparts white, fine dark streaks to flanks and through buffy-breast. Pronounced, broad long white supercilium, dark, streaky crown. In flight, dark brown above, dark barring to white rump, pale brown underwing. Call a thin, fast *chiff-if-if*.

Where to see: Scarce passage migrant from N Europe, winters in Africa. Mostly Autumn Jul–Sep. Rare breeder in Scotland. Freshwater, fringe habitats; also coastal pools and marshes.

Redshank *Tringa totanus* 27–29cm

non-br.

Medium-sized, long-legged wader, with orange-red legs and a straight, medium-length, black-tipped orange bill. Similar to **Spotted Redshank**. Breeding adult and juvenile shows brown upperparts with fine dark barring. Underparts white, densely mottled with fine dark spots and streaks. Non-breeding adult much plainer, grey-brown above and grey-white below. In flight, broad white trailing edge to wing, white tail and long white back wedge conspicuous. Nervous wader, easily flushed. Calls readily, a loud, alarmed *teeu-who-who* or *teu*.

Where to see: Locally common resident, migrant and winter visitor from Iceland and N Europe, Aug–Mar. Favours coastal wetlands and estuaries; saltmarshes in winter. Breeds on coastal wetlands and grasslands mainly in NW.

br.

Spotted Redshank *Tringa erythropus*   29–31cm

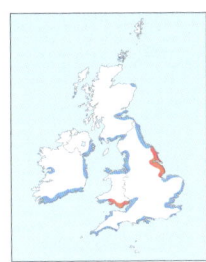

aut.

Elegant, long-legged wader with slim build and fine medium-length bill, which decurves subtly near tip. Breeding adult has striking black plumage with varied white feather fringes to upperparts and flanks. Legs and bill base red. Non-breeding adult plainer greyer above, with pale feather fringes and clean white underparts. Non-breeding adult and juvenile show distinctive white supercilium and dark lores. Juvenile duskier brown throughout with fine barred flanks, like **Redshank**. Elongated white back wedge in flight, but no white trailing edge to wings. Distinct *chui-it* call.

Where to see: Scarce, widespread migrant, Apr–May and Aug-Oct. Breeds in N and NE Europe, winters in S Europe and Africa. Prefers coastal pools, brackish lagoons and estuaries; also freshwater lakes and reservoirs.

non-br.

Greenshank *Tringa nebularia* 30–33cm

br.

Large, robust wader with stout, slightly upcurved bill, overall heavier appearance than **Redshank**. Long legs and bill base both green-grey. Breeding adult brownish-grey above with some dark feather centres, white below with fine dark streaking and spotting to head, neck and upper breast. Non-breeding adult and juvenile plainer grey above, contrasting white feather fringe streaks throughout. Underparts white with reduced, finer dark streaking to head, neck and breast. In flight, uniform brown wings and long, thin white back wedge obvious. Call a strong triple *tew-tew-tew*.

Where to see: Rare, local breeder (N Scotland) Mar–Jul, scarce, widespread migrant and winter visitor Jul–Mar, from Fennoscandia and N Europe. Prefers wetlands both inland and along coast, on estuaries, reservoirs, pools and lagoons.

non-br.

Kittiwake *Rissa tridactyla* 37–42cm

Maritime genuine 'sea-gull', medium-sized, graceful with shallowly forked tail, most similar to **Common Gull**. Breeding adult dapper, pale grey upperparts, pure white head and underparts, bold black 'ink-dipped' wing-tips. Head rounded, bill slender and yellow. Legs blackish. Eye small, dark, red rim at close range. Non-breeding adult similar but soft dark nape collar and ear-spot. Juvenile/1st winter shows dark neck collar, ear-patch, black bill, black tail-band. Upperwing shows bold black 'W' on open wings in flight. Call loud, nasal *kitt-i-waaake* repeated.

Where to see: Locally common, widespread summer visitor and passage migrant, scarcer in winter. Breeds in large coastal cliff colonies, also coastal building ledges, mainly in NW. Winters offshore around the British Isles;, may be seen all year round, most frequently off N coasts.

1st win.

Sabine's Gull *Xema sabini* 30–36cm

1st win.

Small gull. Breeding adult has dark grey hood bordered black. Eye dark, red rimmed, bill small, black with yellow tip. Upperparts dove grey, primaries black with large white tips, underparts white, tail forked. Non-breeding adult has dark nape-band, mottled grey-white head, dark bill. Juvenile brown-fawny above, scaly with buff feather fringes throughout. All ages show triangle patterning to upperwing in flight (from outer edge inwards) wedges of black, white and pale grey (brown in juvenile). All show white tail, juvenile has black tail-band.

Where to see: Rare annual coastal migrant from Arctic, mainly offshore in autumn, Aug–Nov, during passage to South Africa and South America. Mostly off SW coast, occasionally inland after storms.

non-br.

br.

Black-headed Gull *Chroicocephalus ridibundus* 34–37cm

Small, slender, bold gull. Breeding adult has chocolate brown hood, white eye crescent, red bill and bright red legs. Pale grey above, white below, with black primary tips to long, slim wings. Non-breeding adult's hood is reduced to dark ear-spot on white head, bill bright red with black tip. Juvenile shows white-fringed gingery-brown upperparts, and black band across white tail. First-winter like non-breeding adult but browner winged. In flight, all show white wedge to leading edge of wing and black primary tips. Call harsh *krreearr* and sharp *kek*.

Where to see: Common, widespread resident and winter visitor from Iceland and N Europe. Breeds colonially on coasts and inland wetlands. Common on estuaries, farmland, suburban habitats, parks and lakes in winter.

1st win.

non-br.

br.

Little Gull *Hydrocoloeus minutus*   25–27cm

Dainty, small gull with buoyant flight. Breeding adult has black hood, pale grey upperparts, white underparts, often flushed pink. Small dark bill and red legs. Distinctive dusky underwing and white trailing edge to wing visible from above and below. Adult non-breeding shows dark hood replaced with greyish crown and black ear-spot. Juvenile shows white-fringed, brown feathers to upperparts, dusky crown and ear-spot, black tail-tip and black 'W' to upperwing like winter **Kittiwake**. First-winter similar, but upperparts cleaner grey not brown. Call a *repeated kek*.

Where to see: Scarce, locally frequent passage migrant, Jul–Sep, small, variable

1st win.

numbers all year. Found offshore along most of British coast, also estuaries, lakes, reservoirs, even inland. Strong annual spring passage through coastal NE England.

br.

Mediterranean Gull *Ichthyaetus melanocephalus* 36–38cm

br.

1st win.

non-br.

Larger, heavier set than **Black-headed Gull**. Bill deep, chunky, bright red with black-and-yellow tip in breeding adult, contrasting with brown-black full hood and fine white eye crescent. Upperparts pale grey, underparts white. Wings strikingly 'all-white' in flight. Legs bright red. Non-breeding adult has broken, dusky face mask, duller legs. Immature shows variable black markings to primaries, evident in flight and at rest, dark bill and legs. Juvenile shows extensive brown-buff to upperparts, like **Common Gull**. Call distinctive loud, nasal *aw-aw* and *arrr*.

Where to see: Scarce, local resident, locally common passage migrant, winter visitor. Mainly coastal S, E, NW England. Colonial breeder, often with other gull species. Found on coastal lagoons, lakes, reservoirs and docks, often more inland in winter.

Common Gull *Larus canus* 40–42cm

Smaller than **Herring Gull**, soft featured with rounded head. Breeding adult grey above with obvious white spots to primary tips. Head and underparts clean white. Legs yellowy-green, bill yellow, lacking red spot of Herring Gull, eyes dark, rimmed red. Non-breeding adult has grey mottling on white head, grey band through dull green bill. Juvenile buff-fringed brown above, dusky white mottling below. Bill pinkish, dark-tipped. In flight, wing-tips brown-black, black tail-band, white rump. First-winter similar with grey in mantle. Call a high-pitched *mew*.

Where to see: Locally common resident, mainly N and W Scotland, W Ireland. Widespread winter visitor throughout coastal Britain, Sep–Mar. Frequents coastal bays in summer, coasts, wetlands and lakes, and inland areas in winter.

1st win.

non-br.

br.

Great Black-backed Gull *Larus marinus* 64–78cm

br.

1st win.

1st win.

Our largest gull, with sizeable head, chunky bill, thick neck and broad wings. Adult almost black above, white below with large white spots to dark primary tips. Bill deep, yolk yellow with large red lower mandible spot. Eye yellow or greyish, small, with thin red rim. Legs pale, dull pink. Head white all year but with fine grey streaks in winter. Juvenile/ first-winter best identified by sheer size as plumage similar to that of other young gulls. Heavy bill is black, upper-parts rather chequered buff-edged brown, head and underparts paler buff-white.

Where to see: Locally common, widespread resident and winter visitor, Aug–Mar, from N Europe. Mainly coastal, rocky outcrops, moorland to breed, coastal and inland wetlands, farmland, lakes and landfill sites in winter.

Glaucous Gull *Larus hyperboreus* 62–68cm

Large, pale gull, heavy head, rather long, deep bill and, most notable, no black to wings, one of the 'white-winged' gulls. Long primary projection, wings extending beyond tail, head with sloping forehead. Breeding adult pale grey above (paler than **Herring Gull**), white below. Bill restricted red spot to lower mandible. Legs pale pink. Non-breeding adult similar, bill duller, head with grey-brown streaking. Juvenile/first-year variable, pale buff-brown, often pink hued, fine uniform mottling and barring throughout. Bill pale pink, dark tip, eye appears dark. Second-winter shows some pale to buff-brown mottled upperparts, heavy streaking to underparts, whiter face, paler eye.

Where to see: Scarce annual winter visitor, mainly Oct–Apr, from Arctic regions. Found throughout coastal Britain, commonest in N Scottish islands.

1st win.

1st win.

non-br.

Iceland Gull *Larus glaucoides*   52–60cm

Large, **Herring Gull**-sized, 'white-winged gull'. Similar to **Glaucous Gull**, daintier profile, smaller, rounder head, shorter bill, legs. Long white wing-tips extend well past tail. Adult very pale grey above, white head, underparts. Eye straw yellow, bill pale yellow, reduced dull red spot to lower mandible. Non-breeding adult has fine brown streaks on head, neck. Legs pale pink. First-winter pale buff-brown, pink hued, fine uniform mottling, barring throughout, cleaner, paler head. Bill dusky pink, dark tip, becoming yellow-green with dark tip band extending into upper mandible in second and third-winter birds. Second-winter more white, less mottling and streaks to upperparts. Third-winter develops pale grey to upperparts.

Where to see: Scarce annual winter visitor, mainly Oct–Apr, from Arctic regions. Coastal Britain, commonest in N Scottish islands.

1st win.

2nd win.

non-br.

Herring Gull *Larus argentatus* 55–64cm

non-br.

Large, sturdy 'seaside' gull. Breeding adult uniform pale grey above (palest of the large gulls). Wing-tips black with white spots to primaries, and clean white below. Legs pale pink, bill large and yellow with red patch to lower mandible. Non-breeding adult shows heavy streaking to head. Juvenile/first-year variable mottled brown-buff throughout, with pale feather fringes giving scaled appearance. Head cleaner with fine brown streaks and dusky eye. Bill dark. Second-winter has pale grey mantle, making identification easier.

Where to see: Common, widespread resident and winter visitor from NW Europe, Sep–Mar. Mainly coastal but also common inland, frequents harbours, farmland, estuaries, lakes and landfills.

1st win.

2nd win.

Caspian Gull *Larus cachinnans* 57–67cm

1st win.

Large gull, similar to **Herring Gull** and **Yellow-legged Gull** but with longer legs, wings, neck. Comparatively flat forehead, angled crown, longer, parallel-edged bill. Eye appears small, often dark. Adult head white all year. Larger white tips to dark primary feathers. Legs pale, dull green-yellow to pale pink. Bill green-yellow, dull red lower mandible spot, becoming dark bar into upper mandible in winter. Juvenile and first-winter uniform, mottled brown and buff, greying back. Lacks wavy patterning of buff-pale feather fringes shown in Herring Gull tertials. Often whiter head and under-parts than Herring Gull.

Where to see: Rare, increasing winter visitor from E Europe, Aug–Apr. Found with other large gulls along coast, mainly SE England. Also inland lakes, landfill sites and gull roosts.

Yellow-legged Gull *Larus michahellis* 55–64cm

1st win.

The continental counterpart to **Herring Gull**. Breeding adult differs in having yellow legs, smaller white spots on black outer primaries, larger red spot on lower mandible often extending onto upper mandible, and greyer upperparts, though still paler grey than **Lesser Black-backed Gull**. Non-breeding adult has very restricted streaking around white head, the white head standing out in flocks of winter Herring Gulls. Juvenile/first-winter

difficult, but often warmer brown-toned, grey to hindneck, paler-headed with dusky smudge around eyes, whitish rump and tail and blackish tail-band.

Where to see: Scarce but regular annual migrant from S Europe, mainly Aug–Apr. Rare breeder. Mainly found in England, along coasts and around lakes and reservoirs, including inland.

Lesser Black-backed Gull *Larus fuscus* 52–64cm

1st win.

Large, slim, long-winged gull. Breeding adult has slate-grey upperparts, pure white underparts, contrasting black primary tips, bright yellow legs. Large yellow bill, red patch on lower mandible. Eyes pale, rimmed red. Non-breeding adult shows dusky streaks around head. Juvenile/first-winter mottled grey-brown, buff-fringed dark brown feathers above, streaky and mottled buff-brown below. Paler, cleaner head and neck, dusky around eye. Bill dark, legs pink. Second-winter grey to mantle, paler underparts. Third-winter more like adult, grey to mantle and wings, some mottling to head and breast.

Where to see: Common, widespread resident, summer visitor from North Africa, Apr–Aug, winter visitor NW Europe, Sep–Mar. Around most coasts all year. Colonial coastal and inland breeder, including marshes, moorland, islands and even localised or urban roofs.

br.

Sandwich Tern *Thalasseus sandvicensis* 36–41cm

Large, very white tern. Distinctive front-heavy appearance with large head, long slender bill, long pointed wings and short forked tail. Breeding adult has yellow tip to black bill, black crown with spiky crest and short black legs. Upperparts pale grey with diffused black wedge to outer wing primaries. Underparts white. Non-breeding adult (from July) has white forehead and restricted rear black crown, giving masked appearance. Juvenile has dusky speckled cap, all-black bill, dark scaly feather fringes on grey upperparts, and dark outer wing. Call a grating *kirreet*.

Where to see: Locally common and widespread summer visitor, Mar–Oct, winters in Africa. Breeds in scattered coastal areas around the British Isles in colonies on sand and shingle beaches.

juv.

non-br.

br.

Little Tern *Sternula albifrons* 22–24cm

Small, heavy-fronted tern; narrow wings, large head, short forked tail. Pale grey above, white below, black crown extends to nape, black eye-stripe, white forehead, and cheek. Bill yellow, black tipped, legs short, orange. Rump and tail white, wing with black leading edge. Juvenile has dusky crown, grey-brown scaly above, white below. Flight direct, hovers and plunge dives for food. Call a sharp *kreet*.

Where to see: Scarce, local coastal summer visitor, Apr–Sep, winters in Africa. Shingle and sandy beaches, loose colonial nester. Mainly breeds E England, widespread offshore migrant.

Roseate Tern *Sterna dougallii* 33–36cm

Silver-grey above, white below, often tinged pink. Adult has black cap, extending to nape; black bill with reddish base in summer, red legs. Thin black leading edge to wing, long white tail-streamers. In flight, black diffused wedge to upper outer wing like **Common Tern** and **Sandwich Tern**. Juvenile shows dusky forehead, black bill and legs.

Where to see: Rare, very local summer visitor, Apr–Oct, rare passage migrant, winters in Africa. Coasts, offshore islands, beaches. Coquet Island in Northumberland is the only regular breeding location in British Isles. More widespread during migration.

Common Tern *Sterna hirundo* 34–37cm

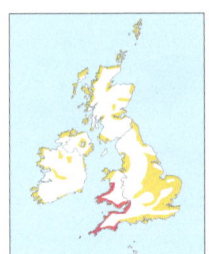

Medium-sized, slender tern with forked tail and short tail-streamers. Breeding adult has pale grey upperparts, whitish underparts, black-tipped bright red bill, short red legs, and smart black cap extending onto nape. Wings long and slender, with darker grey primaries towards tips. Juvenile has buff-white forehead, black rear crown mask and darkening carpal joint bar. Upperparts scaly with ginger-buff fringes to pale feathers, underparts buffy-white, and bill orange with dark tip.

Where to see: Locally common, widespread summer visitor and passage migrant, Apr–Oct, winters in Africa. Scattered colonial breeder along coast and inland. Nests on or around beaches, rocky outcrops, artificial rafts, reservoirs, lakes and rivers.

juv.

Arctic Tern *Sterna paradisaea* 33–39cm

juv.

Like **Common Tern** but paler grey above, bright all-red, shorter bill, shorter red legs, much longer tail-streamers. Breeding adult silver-grey above, off-white below, greyer than Common with similar clean black cap. In flight shows pale grey wings with dark trailing edge, otherwise rather uniform; Common has broader trailing edge, diffused darker grey wedge to outer primaries. Juvenile similar to Common but has greyer upperparts, redder bill base, distinct white trailing edge to inner wing in flight.

Where to see: Locally common summer visitor, Apr–Oct, mostly Scotland and Ireland, widespread coastal and inland passage migrant. Longest migration of any bird species, winters in Antarctica. Prefers coasts, offshore islands to breed, may be seen at inland and coastal wetlands or offshore during migration.

Black Tern *Chlidonias niger* 22–24cm

Small 'marsh' tern, with short, broad wings and short, shallow-forked tail. Breeding adult dark grey above, sooty-black below from head to underbelly contrasting with white undertail. Bill short and black, legs dusky red. Non-breeding adult replaces sooty underparts and head with white, black reduced to encompass cap, nape and ear-coverts. White collar with dusky shoulder-patch. Juvenile like winter adult but with buff-edged brown scaly feathers to back and wings.

Where to see: Scarce passage spring and autumn migrant, Apr–Oct, breeds in NW Europe, winters in Africa. Migrants use inland lakes and reservoirs, also coastal lagoons and marshes; mainly England.

Great Skua *Stercorarius skua*   50–58cm

Large, powerful, stocky skua, heavy chested with short dark tail, legs and chunky black bill. Dark brown throughout, flecked with variable pale-buff streaks.

Juvenile/first-winter less marked and darker brown. Obvious large white 'comma' flash in outer wings from above and below. Can be aggressive when on territory. Displays on ground with wings raised, or stoops, stiff-winged in mid-air crowing a nasal *arh-arh-arh* while descending. Steals food from other seabirds, including Gannets.

Where to see: Scarce, local, summer visitor, Apr–Nov, from South Atlantic and Brazil and uncommon, widespread offshore passage migrant. Breeds along coastal W Ireland, N Scotland, Northern Isles. Offshore during mainly autumn passage along wider British coasts, especially after storms.

Pomarine Skua *Stercorarius pomarinus*   38–57cm

Bulky, heavy-chested, long broad wings, long spoon-like central tail feathers (sometimes absent/damaged). Adult has dark brown upperparts, crown, face mask.

Underparts white, head and neck creamy; with yellow hues, variable brown breast–band (sometimes absent), dark barred flanks. Rarer dark morph adult brown throughout. Juvenile brown, fine white barred underparts. All show white 'comma' flash to outer-wing above, and split, double white outer-wing patch below.

Where to see: Scarce, widespread coastal passage migrant, Apr–May and Aug–Nov. Breeds in Siberia, winters in West Africa. Mostly small groups offshore after gales.

Arctic Skua *Stercorarius parasiticus*   41–46cm

Graceful, slender, with pointed wings, long pointed central tail feathers, white 'flashes' to outer wings above and below. Two colour morphs. Dark morph chocolate-brown throughout. Pale morph has dark brown cap and upperparts, pale buff-cream face, collar and underparts, dusky broken breast-band. Juvenile mottled buff-edged brown throughout, tail shorter. Chases terns and other seabirds to steal food, with agile, powerful flight. Call harsh, nasal mewing.

Where to see: Scarce, local summer visitor Apr–Nov, from S Atlantic, uncommon passage migrant. Declining breeder, coastal N and W Scotland, Northern Isles. More widespread offshore during autumn migration, especially after storms.

juv.

pale morph

dark morph

Long-tailed Skua *Stercorarius longicaudus* 38–57cm

Relatively small, slender, long, pointed wings. Tail long, rounded with very long, pointed central tail feathers. Adult grey-brown above, white below with dusky belly. Blackish cap on cream-yellow head, white breast, no breast-band. Restricted white flashes on outer-wing above and below. Dark morph adult; all brown, very rare. Juvenile brown, with pale-buff feather edges, paler head, belly. Tail tapering but blunt.

Where to see: Rare, offshore passage migrant, mainly off NW; May–Jun and off E; Aug–Oct. Occasionally good numbers during spring passage. Breeds in Fennoscandia, N Russia; winters in S Atlantic.

Little Auk *Alle alle* 19–21cm

Smallest auk, size of Starling, pied plumage. Dumpy, short neck, wings and tail, stubby dark bill. Breeding adult rare; dark glossy brown head, neck, upper breast, pure white lower breast and underparts. Adult non-breeding/first-winter shows white extending to throat and around collar. Upperparts; hood, forehead and eye-surround, appear black not brown. In flight, underwing dark.

Where to see: Scarce, widespread autumn and winter migrant, mainly offshore, sometimes close to shore after gales. Breeds in high Arctic, Greenland and Spitsbergen. Most off N and E coast of British Isles.

non-br.

Common Guillemot *Uria aalge* 38–46cm

br.

Similar pied plumage to **Razorbill** but more chocolate brown than black, with smooth, oval-shaped head, longer, slender dark bill and neck, and blunt tail. White underparts broken by dark flank streaking. White wing-bar at rest. A few are of the 'bridled' form, with fine white eye-ring extending and curving down into cheek on otherwise uniform dark brown upperparts. Slender profile and upright 'penguin' stance on land. Non-breeding adult and juvenile white through upper breast, throat, cheek. White cheek broken by narrow black line extending from eye.

Where to see: Locally common, widespread resident, found in summer months on sea cliffs and islands; mainly N and W, absent from SE. Colonial nester, breeding in large tightly packed groups on open ledges. Winters offshore.

non-br.

bridled form

Razorbill *Alca torda* 38–43cm

1st win.

Stocky, thick-necked black-and-white auk with deep, vertically flattened, blunt-ended bill and pointed tail. Body rather heavy, wings and legs short. Breeding adult uniform black above, white below. Fine white line horizontally from upper bill base to eye and vertically through black bill near tip. Non-breeding adult and juvenile similar, but extensive white to upper breast and throat, diffused smoky white cheek-patch and no white face line. All show narrow white wing-bar at rest.

Where to see: Locally common, wide-spread resident, breeds mainly on N and W coast, absent from SE. Summer cliff nester prefers rock crevices and boulders than open ledges. Loosely colonial, often scattered pairs close to and among bigger **Guillemot** colonies. Winters offshore.

br.

Black Guillemot *Cepphus grylle*   32–38cm

non-br.

Structurally like small **Guillemot**. Breeding adult distinctive, dark chocolate brown throughout, large rounded white wing-patch, bright red legs and red inside dark bill (only seen when bill is open). In flight, shows white underwing. Non-breeding adult very different, mottled and barred black, grey and white plumage. Shows black wing with large white wing-patch in flight. Bill and legs dark. Juvenile similar but with less defined white wing-patch.

Where to see: Fairly common, local resident, mainly coastal Scotland and Ireland. Winters just offshore but often around harbours and cliff-bases, close to summer breeding sites. Nests in pairs or loose colonies, generally out of sight in low cliff crevices and caves around rocky coasts and islands.

br.

Puffin *Fratercula arctica*   28–34cm

Rotund, small auk with upright stance, unique breeding plumage and charismatic behaviour. Breeding adult black above, white below. Head rounded, face white bordered black and with large triangular red, blue-grey and yellow bill. Eye surround has red and grey coloured triangular skin-patches. Legs short, bright orange. Non-breeding adult becomes dusky, dirty-faced, bill becomes smaller, grey-blue and red. Juvenile has soot-dipped, dusky face. Eye, legs and bill all dark, bill much narrower than non-breeding adult.

Where to see: Locally common coastal summer resident, mainly N and W Scotland but also Ireland, Wales and SW and NE England. Nests in burrows and crevices within clifftop colonies and coastal slopes along rocky coastlines and offshore islands. Winters offshore.

Red-throated Diver *Gavia stellata* 55–67cm

non-br.

Commonest, **Mallard**-sized diver. Long body, slim, upright neck and slender, slightly upturned, tapering bill. Breeding adult dark grey-brown above, white below. Head and neck storm grey with fine white and black lines running up nape, and large brick red throat-patch that can appear dark in poor light. Eye red, bill dark grey. Non-breeding adult/first-winter grey above with fine white flecks. White below extends through breast, throat, lower head, almost to hindneck and surrounds beady dark eye. Grey crown narrow, extending through softly sloping forehead and hindneck.

Where to see: Scarce, local breeder, mainly inland freshwater lochs and pools close to the sea, NW Scotland, Scottish islands. Scarce winter visitor and passage migrant offshore and inshore around all coastal UK.

br.

Black-throated Diver *Gavia arctica* 63–75cm

non-br.

Larger than **Red-throated Diver**, with thicker neck and straighter bill. Head rounded, forehead steep. Breeding adult black above with white chequers along back. Head and nape velvet dark grey, with black throat-patch surrounded by fine, tight black-and-white stripes, wider through neck, denser on breast. Bill and eyes dark. Non-breeding adult/first-winter plainer grey, black above, lighter, smoky brown-black through nape, neck-sides and head, encompassing eye. Bill dull grey. Underparts white, well-defined through neck, throat, into lower face and cheek. First-winter shows buff-white scaly feather fringes along back. All show white patch to hindflank at rest.

Where to see: Rare, local breeder, large lakes in N Scotland. Rare winter visitor and passage migrant offshore all UK.

br.

Great Northern Diver *Gavia immer* 73–88cm

Largest diver, stocky with thick neck, large head with steep forehead and dagger-shaped bill. Breeding adult black above with extensive white chequering and spots. Head and neck sooty-black, broken by fine black-and-white striped triangular neck-patches. White breast with black-and-white stripes along sides. Eye red, bill black. Non-breeding adult and first-winter told from other divers at distance by chunky size, large head, bill and steep 'bumpy' forehead. Nape, hindneck and cap blackish, as dark as back. Bill paler grey in winter.

Where to see: Scarce winter visitor, Aug–May, offshore, also inshore in sheltered bays and estuaries. Breeds in Iceland and Greenland.

non-br.

br.

Storm Petrel *Hydrobates pelagicus* 15–16cm

Tiny, flutters and glides low over sea feeding, patters surface with feet down, wings raised. Sooty black with white rump, white bar across central underwing. Wings fairly broad, stiff, softly curved, round ended. Tail short, square ended. Tiny black bill, steep forehead.

Where to see: Scarce, locally common summer visitor, Apr–Oct, scarce, widespread offshore passage migrant, winters off Africa. Colonial coastal breeder, offshore islands, N and W. Nests in crevices and burrows along rocky shores, boulder beaches, stone walls. Visits colonies at night to avoid predators. Mostly seen offshore.

Leach's Petrel *Oceanodroma leucorhoa* 18–21cm

Small, but larger than **Storm Petrel**, similar open-ocean surface feeder, more tern-like in behaviour. Sooty black throughout, longer, angular, pointed wings. Upperwing with broad, grey arched band, shallowly forked tail, narrow white rump. Underwing uniformly dark.

Where to see: Very scarce, local summer visitor, scarce passage migrant, Apr–Nov; winters in the tropics, S Atlantic. Colonial breeder, on Scottish and Irish offshore islands, only attended at night.

On migration offshore W Scotland, W and SW England and Wales. After strong NW or W gales can be close inshore, NW England.

Fulmar *Fulmarus glacialis* 43–52cm

Large stocky petrel, gull-like with variable pale grey upperparts, pale patch on outer wing, and white head and underparts. Neck very thick, bill blue-yellow, chunky, showing 'tube-nose' nostrils halfway along upper mandible and lightly hooked, domed orange-yellow bill tip. Dark eye with smoky-grey surround. Strong flap and glide flight on stiff wings. Readily swims with head high, tail raised, but cumbersome on land. Scarce 'northern' forms are variably darker, smoky throughout.

Where to see: Locally common, widespread cliff breeder throughout coastal UK, Mar–Sep. Offshore and around coastal cliffs all year. Nests in loose colonies dotted around exposed cliffs.

Manx Shearwater *Puffinus puffinus* 30–35cm

Commonest shearwater in UK waters. Slender, tapered body, long, straight wings held stiffly in flight. Dark brown above, appears black at distance, extending along nape, head and encompassing eye area. Extensively white through flanks, belly, breast, throat into lower face and cheek. Underwing white, bordered black. Bill black, narrow, rather long with hooked tip. Wingbeats fast, followed by strong glides and high banking, black-and-white flashes as it tilts over waves. Forms 'rafts' on water.

Where to see: Locally common coastal summer visitor, common, widespread offshore passage migrant, Mar–Oct, winters in South America. Breeds in burrow colonies on a few offshore islands, N and W coast Scotland, Ireland and Wales. May gather in evening rafts close offshore, attends burrows at night. Commonly seen offshore during migration.

Sooty Shearwater *Ardenna grisea*   40–50cm

Scarcer, larger, heavier-bodied than **Manx Shearwater**, all dark smoky-brown plumage. Dark brown above and below broken by broad, grey-white, diffused band along underwing, evident even at distance. Agile in flight, wings appear swept back, with a few slow yet strong wingbeats, and long 'shearing' glides low over water.

Where to see: Scarce, regular passage migrant, offshore Jul–Oct, all UK. Migrates from N Atlantic to S oceans and Pacific to breed. Best seen at sea or from headland sea watches.

Gannet *Morus bassanus* 85–97cm

juv.

Very large, robust seabird with chunky, rounded head and neck, dagger-like bill, and long, angled wings. Adult pure white with black wing-tips and light yellow-orange head and nape. Bill deep, pale grey with fine black lines. Eyes pale blue with blue-grey eye-ring neatly bordered black. Legs short, dark brown with green stripes. Plunge-dives for fish. Flight straight, measured, alone or in lines. Juvenile all brown with white spotting throughout.

Immature (under four years) shows extensive black-brown to upperparts, decreasing with age.

Where to see: Locally common resident, on breeding sites Mar–Oct. Frequent, widespread offshore migrant, most coasts all year. Colonial nester, often many thousands of birds, on established cliff sites, mainly offshore islands, N and W Britain and Ireland.

Shag *Gulosus aristotelis* 68–78cm

Smaller than **Cormorant**, with smaller, rounded head and steep forehead. Breeding adult blackish throughout, with glossy green sheen and dark-edged 'scaling' to wing feathers. Prominent yellow 'gape' at bill base and short, erect dark crest often visible. Eye emerald-green, bill and legs dark. Non-breeding adult duller. Juvenile brown with variable washed-out, dull brown underparts.

White chin-spot and yellow-grey bill. Low water profile, head tilted upwards. Upright on land, often seen drying outstretched wings.

Where to see: Fairly common resident, mainly N and W coastal British Isles. Smaller numbers along all coasts in winter. Nest on cliffs, caves, boulder beaches, rarely inland.

juv.

Cormorant *Phalacrocorax carbo* 77–94cm

Larger, bulkier than **Shag** with elongated body, long, slim neck and rather long tail. Long, straight, hook-tipped bill and sloping forehead, peaking at rear. Breeding adult black with bronze-blue sheen, wings with dark-edged feathering, giving scaly appearance. Yellowy skin-patch at bill base, white throat and cheek-patch. White hindflank-patch present only when breeding. Non-breeding adult duller, white areas around face duskier, less defined. Juvenile dull brown, underparts off-white. Swims low in water, dives for fish. On land stands very upright or awkwardly horizontal.

Where to see: Common coastal resident throughout the British Isles, also winter

visitor from N Europe. Loose colonial breeder on rocky shores and cliffs. Also inland waters, nesting in waterside trees.

Glossy Ibis *Plegadis falcinellus* 55–65cm

Large, elegant waterbird, resembling a large black **Curlew** at distance, with long, downcurved, sickle-shaped bill, proportionally short legs, and narrow, long neck. Breeding adult has reddish purple-green gloss throughout, white area at bill base and variable dark pinkish-grey bill and legs. Non-breeding adult and juvenile duller bronzy-brown throughout, lacking rich purple gloss, with fine white flecks to head and neck. Flight strong, legs and neck fully extended, often in groups and line formations.

Where to see: Very scarce migrant and winter visitor from SW Europe (successfully bred in 2022). Mostly immature birds in autumn in S coastal England and Ireland. Prefers wet fields, marshes, lagoon fringes, estuaries and bays.

White Stork *Ciconia ciconia* 95–110cm

Huge white waterbird. Often appears dirty white, with long, black-tipped wings, long neck, large dagger-shaped orange-red bill, and sturdy red-pink legs. In flight, broad wings show thick black band across outer trailing edge and obvious 'fingered' wingtips, legs project well past tail. Often glides, sometimes to great height on thermals. Juvenile duller throughout, black bill. Confidently walks and hunts on foot.

Where to see: Very scarce migrant, from S Europe, mostly May–Oct, S and E England, winters in South Africa. Wet fields, grasslands, marshes, ditches.

Spoonbill *Platalea leucorodia* 80–93cm

Large, robust yet elegant white waterbird, with a unique, long, spoon-shaped bill, long neck and legs. Breeding adult has a yellowish tinge to shoulders and short, square-ended crest plumes. Legs and bill black, the latter tipped yellow. Non-breeding adult all-white. Juvenile similar but with black tips to wings, evident in flight, grey legs and pinkish spatulate bill. Flies with characteristic fast wingbeats, with legs extended and neck fully outstretched, unlike similar **Great White Egret**.

Where to see: Very rare local coastal resident, scarce widespread migrant and winter visitor from SW Europe. Breeds sparsely in wetland trees, S and E coast England. Frequents coastal inlets, marshes, estuaries and river valleys.

Bittern *Botaurus stellaris* 69–81cm

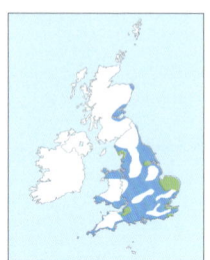

Large, stocky, rather hunched brown heron, thick neck, short legs and prominent bill. Very secretive nature. Adult pale buff to tawny-brown plumage with black-and-brown mottling and bars throughout. Blackish markings boldest along back. Head with black crown and moustachial stripe. Breast with heavy black vertical stripes. Bill and legs chunky, yellowish-green. Juvenile similar. Call a characteristic hollow, low boom, made by territorial male. Can be almost invisible in reedbed with cryptic plumage and statue stance, often with head sky-pointing.

Where to see: Rare local resident, scarce, widespread winter visitor from N Europe. Breeds mainly in England, more widespread into Wales and Scotland in winter. Prefers freshwater, coastal extensive reedbeds, also other wetland marshes, river fringes and ditches.

Cattle Egret *Bubulcus ibis* 45–52cm

Small white waterbird, not as slim as **Little Egret**, with more rounded features, stouter, shorter dagger-like bill, shorter legs and puffy throat. Adult summer white with yellow bill often, variable reddish base and warm buff-orange through short-plumed crown, hindneck, breast and back. Legs paler straw yellow-brown. Adult winter and juvenile lack orange plumage and plumes, being all white, with pale yellow bill and duller yellow-brown legs.

Where to see: Rare resident and scarce winter visitor from S Europe, Oct–Mar. Breeds in England and Wales. Mainly coastal but expanding inland, favours wet grasslands, marshes, ditches and fields, often around livestock; also coastal lagoons and lake fringes. More widespread, SW bias in winter.

Grey Heron *Ardea cinerea*   84–102cm

Large, robust heron, slender body, long legs and neck, heavy dagger-like bill, grey-and-white plumage. Adult has white head, neck and underparts, bold black plume-stripe behind eye, fine black streaks through neck and breast. Bill yellow, more orange in summer. Upperparts storm grey, legs yellow-grey. Juvenile duller, with grey cap, no black plume stripe behind eye and grey-yellow bill. In flight, head and neck retract towards body, creating a bulge, legs extend well beyond tail. Broad wings show black to all outer half and inner trailing edge. May stand statue still for long periods, either hunched at rest or poised, ready to lunge at prey.

Where to see: Common, widespread wetland resident and winter visitor from N Europe. Favours fringe habitats both coastal and freshwater.

Purple Heron *Ardea purpurea* 70–90cm

Smaller than **Grey Heron**, with narrow head, long, slender neck and long, pointed bill. Adult upperparts grey-mauve with purple-red shoulder-patch, brown back plumes and blackish flanks. Head, neck and breast with rufous, black and cream stripes and streaks. Cap black with black nape-plumes. Bill and legs yellow-orange. Juvenile chestnut brown, mottled, with paler orange-buff with black streaks through neck and breast. Face with black-and-white stripes. Deep bend to neck in flight, legs fully extend past tail. First-summer like adult but brown-winged.

Where to see: Rare migrant from S Europe, winters in Africa. Mainly adults seen in S England, Apr–Aug. Has bred. Prefers shallow wetlands, coastal reeds and marshes.

Great White Egret *Ardea alba*   85–100cm

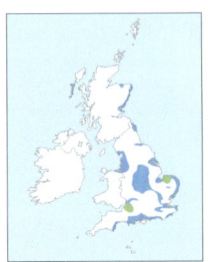

Large white waterbird, size of **Grey Heron**, with a long, often 'kinked' neck, long legs and long, straight, dagger-like bill. Legs are dark, not yellow-footed like **Little Egret**. Breeding adult has all-dark bill and fine white plumes only through upper breast and along lower back, extending past tail. Non-breeding adult and juvenile show striking all-yellow bill and variable yellow to upper legs. Flight rather lethargic, with head and neck retracted and long legs extending far beyond tail.

Where to see: Scarce, annual migrant and winter visitor to coastal S and E England, from Europe. Increasing but rare breeder. Prefers shallow wetlands, both fresh and saltwater, from estuaries and reedbeds to rivers and lake fringes.

Little Egret *Egretta garzetta* 55–65cm

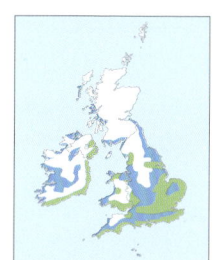

non-br.

br.

Small, slender white waterbird with long, thin neck, broad white wings, long black legs and yellow feet. Bill straight, long, dark grey-black. Bare skin at bill base variable, pink-blue to grey, often yellow in spring. Breeding adult has two fine, long white plumes from hindcrown, shorter plumes around breast and lower back. Non-breeding adult and juvenile lack plumes. Juvenile often shows yellow through lower mandible.

Where to see: Increasingly common resident and winter visitor of freshwater and coastal waters, estuaries, lakes, rivers, wet grasslands and saltmarshes with shallow waters. Expanding range northwards from southern stronghold, and increasingly found inland. Feeds alone, breeds in colonies, often in trees with **Grey Heron**.

Osprey *Pandion haliaetus* 55–58cm

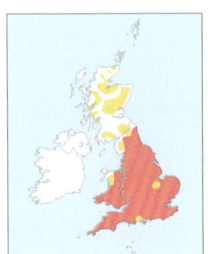

Distinctive large, long-winged, boldly brown-and-white raptor. Adult dark brown above, white below, brown mask through white head, dusky, diffused necklace. Eyes yellow, bill grey, legs feathered white. Juvenile shows scaly pale-fringed brown upperparts. In flight, from above, all brown but for white head and brown-and-white chequered tail. From below, wings and tail finely chequered brown-and-white, dark brown carpal patches, brown curved bar along central underwing. Black, deeply fingered wing-tips, dark band at tail end. Readily hovers and plummets into water to grab fish prey with talons.

Where to see: Scarce summer visitor, scarce, widespread passage migrant, Mar–Oct, winters in Africa. Breeding stronghold in Scotland; also Wales, reintroduced into England. Prefers lakes with adjacent forest, migrants use open wetlands, estuaries, lakes and rivers.

Honey-buzzard *Pernis apivorus*   52–60cm

♂

Large, shy raptor, like **Common Buzzard** in structure but longer, narrower wings and tail, and slender, longer neck. Variable plumage. Adult male shows pale grey head, brown-grey upperparts and clean white underparts with bold spotting. Tail with black band at tip and two more near base, from above and below, weaker on female. Female browner through head, mottled brown-white underparts with streaking rather than spots. Both show round yellow eye and legs. Juvenile variable, can show pale head with dark mask. In flight, all show distinctive projected neck and tail with clear, well-spaced bands, not finely barred like Common Buzzard.

Where to see: Rare summer visitor from Africa, May–Oct, also scarce spring and autumn passage migrant mainly along E coast UK. Elusive woodland species, prefers broad-leaved but also coniferous woodlands, valleys and parklands.

♀

Golden Eagle *Aquila chrysaetos* 75–88cm

Very large, powerful dark brown eagle with long, broad tail and wings. Adult shows conspicuous paler golden-brown 'shawl' around head and nape. Variable bleached brown wing-panel, and variable mottled brown through upperparts and dark brown and greyish tail. Chunky bill grey with yellow base, legs yellow, heavily feathered. Juvenile darker than adult, with variable white on flight feathers, forming white bar or patch on underwing, and white base to black banded tail, most obvious in flight. Develops adult plumage over several years. Wings often raised into 'V' when soaring.

Where to see: Local, rare resident in Scotland, introduced in Northern Ireland, occasionally strays elsewhere. Prefers open wild terrain, old forests, rugged high mountains and island cliffs.

juv.

Birds of prey

Sparrowhawk *Accipiter nisus*   28–38cm

Small raptor with rounded wings and long, square-ended tail. Slender male has slate-grey head and upperparts, with variable white patches. Underparts white, with peachy-orange wash to cheek, throat and flanks, and orange-brown barring through breast. Thin legs yellow, eyes yellow to orange. Female and juvenile dull grey-brown above with yellow eyes, white eyebrow stripe, and white below, with dense warm brown barring on female, and broken brown barring and heart-shaped spots on juvenile. Resembles small male **Goshawk**, but smaller and slighter. All show strong, well-spaced dark tail bars. Very agile hunter.

♂

Where to see: Fairly common resident, breeds in woodland. Frequents forests, farmland, parks and gardens, and urban areas.

♀ juv.

Goshawk *Accipiter gentilis* 48–62cm

Large, strong raptor, with broad wings and long, broadly banded tail. Female substantially larger than male. Male may be confused with female **Sparrowhawk** in flight, but Goshawk larger, heavier chested, with big head, rounded tail and broader wings. Adult male slate grey above, white with dense dark barring below. Face with white eyebrow stripe and broad dark grey cheek-patch. Orange eyes, yellow legs. Conspicuous white undertail 'vent' at rest. Adult female and juvenile browner above. Juvenile warm buff below with dark vertical streaking.

Where to see: Scarce, local resident with scattered distribution throughout. Stronghold Wales, S Scotland and Ireland.

Very elusive mature woodland species, also hunts in adjacent open countryside, heaths and farmland.

Marsh Harrier *Circus aeruginosus* 48–56cm

Large, narrow-winged harrier, the size of **Common Buzzard** but slimmer with long wings and tail, and notably long legs. Often flies with wings held in a distinct 'V'. Male strikingly tricoloured, with pale grey wings contrasting with chestnut back, shoulders and body, and large black tips to primaries. Legs and bill base yellow. Female and juvenile plainer brown throughout, often with cream-buff crown, throat and shoulders. Female larger than male.

Where to see: Scarce, local resident, scarce, more widespread passage migrant and winter visitor from NW Europe. Prefers coastal reedbeds, marshes, open wetlands and farmland.

Hen Harrier *Circus cyaneus*   44–52cm

♂

Smaller than **Marsh Harrier**, with small head and long wings and tail. Male has pale grey head and upperparts, white rump, clear-cut white underparts and a large black primary-tip wedge, evident from above and below. In flight, shows five wing-tip 'fingers' and dark grey trailing edge to underwing. Eyes, bill base and legs bright yellow. Female and juvenile show brown upperparts, pale facial disc and broadly banded tail. Underparts warm brown and white, with variable brown streaking and dark-barred wings (like female **Montagu's Harrier**). Juvenile underparts warm buff, heavily streaked, unlike plainer juvenile Montagu's.

Where to see: Scarce resident and passage migrant from N and W Europe. Breeds in upland heather moorland in NW Scotland and England, N Wales, south-central Ireland. Favours open lowland wetlands, marshes, reedbeds and farmland at other times.

♀

Montagu's Harrier *Circus pygargus* 43–47cm

Small harrier, with slender body, long, narrow wings and tail. Adult male grey above with paler grey rump and black wing-tip wedge. Shows black wing-patch and long wing-tips at rest. Head and breast grey, belly white with fine chestnut streaks. In flight, four 'fingers' on black wing-tips and black bar across upperwing. Underwing with rufous spotting and two black bars. Adult female brown above, buff below with extensive dark streaks. Whitish eye-mask and brown cheek-patch. Bill grey, eyes and legs yellow. Juvenile brown above, unstreaked rufous-orange below. Strong eye-mask and cheek-patch.

Where to see: Rare, local summer visitor, Apr–Sep, rare migrant, winters in Africa. Has bred in S and E England. Prefers open farmland, moors, heaths, grassland and coastal marshes.

Red Kite *Milvus milvus* 60–66cm

Large, long-winged raptor with distinctive long, deeply forked tail. Adult shows striking warm rufous-brown upperparts with black centred feathering throughout. Head pale grey with fine black streaks. Underparts rufous-brown, with dark streaks, bill and legs yellow. Juvenile variable, with pale feather fringes and less forked tail. In flight, tail bright orange-brown above, whiter below, with fine dark barring. Upperwing with broad pale band across blackish wing. Underwing grey-brown with distinct white 'window' patches to outer wing and black fingered tips. Aerobatic, slow, twisting flight, often using tail as a rudder.

Where to see: Locally common, increasing resident with patchy, mostly southern distribution. Prefers deciduous woodland to breed, also open countryside, farmland, roadsides and species-specific feeding stations.

White-tailed Eagle *Haliaeetus albicilla* 70–90cm

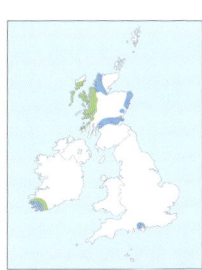

Huge eagle with long, broad 'barn door' wings, and rather short, wedge-shaped tail. Adult cool brown with variable, whitish head, neck and feather fringes through upperparts. Tail white, powerful bill and legs yellow, the latter partly obscured by heavily feathered brown 'trousers'. Juvenile uniformly darker warm brown, including head, neck and tail, with blackish streaks to paler breast and underparts. Bill duller grey with paler base. Develops adult plumage over several years. Flatter winged than **Golden Eagle** in flight, without 'V' profile.

Where to see: Rare, reintroduced resident, mostly NW Scotland, also SW Ireland and S coast of England and Isle of Wight. Very scarce passage migrant, mainly along E coast UK. Frequents remote coastal cliffs, islands and large lakes, also old forests.

Rough-legged Buzzard *Buteo lagopus* 49–59cm

Like **Common Buzzard**, but larger, with longer wings and tail. Adult grey-brown above, with whitish head, breast and underparts, flecked brown, and a bold, variable dark brown belly. In flight, underparts and undertail mainly pale, whitish with contrasting dark brown belly, dark brown carpal joint patches (see Common Buzzard) and narrow, dark trailing edge to outer wing, wing and tail-tips.

In flight from above, whitish outer wing-patches and white tail with dark band (see Common Buzzard). Female similar, juvenile often whiter-headed than adult.

Where to see: Rare passage migrant and winter visitor from Fennoscandia, Oct–Mar, along E coast UK, from N Scotland to S England. Found along open coastal wetlands, marshes and farmland.

Common Buzzard *Buteo buteo* 51–57cm

Large, compact raptor with rounded head, broad wings and short tail. Adult plumage variable, most are brown above, white-brown mottled below with paler belly-crescent. In flight, underparts well marked with head, body and forewing to carpal joint dark brown, with clear contrast against white, finely barred central underwing and dark brown trailing edge to wing and tail-tip. Eyes dark, bill base and legs yellow.

Juvenile streaky on belly, and with less defined dark trailing edge to underwing and tail. Often soars and glides over terrain on stiff, slightly raised wings. Also perches low in trees and on fence posts.

Where to see: Common resident. Nests in woodland or craggy slopes, hunts over woodland, heaths, moorland, farmland and hills.

Kestrel *Falco tinnunculus*   32–35cm

Small, long-tailed falcon with hovering hunting technique. Adult male has grey head, dark moustachial stripe. Chestnut brown above, warm buff below, black spotting throughout. Wing-tips and tail grey-black at rest. In flight, from above, male shows chestnut back and inner wing, black outer wing, grey tail with black band at tip. Female and juvenile warm brown above, strong black bars, buff-white below with dark streaks. Head buff-white with warmer crown and dark moustachial stripe. In flight, pale brown above with darker outer wing, grey rump, barred tail with black band at tip. All show grey-yellow bill and yellow legs.

Where to see: Fairly common, widespread resident. Prefers open country, heaths, roadside verges, suburban habitats and farmland. Nests in buildings or trees.

Merlin *Falco columbarius* 25–30cm

Small, fast, compact falcon with short, pointed wings. Male grey-blue above, orange-buff below and around collar, fine black streaking throughout. Head grey and orange-buff, weak moustachial stripe. Long tail grey, black band at tip. In flight, from above, grey-blue with dark outer wing and tail-tip. Female and juvenile like **Kestrel**, but cooler grey-brown above, tail brown with cream bars, underparts buff-white with blotchy brown streaks. Crown, cheek-patch, moustachial stripe brown. Fast aerial hunter, low over terrain in pursuit of small birds or perched briefly on fence posts or rocky outcrops.

Where to see: Scarce local resident, scarce winter visitor from Iceland, Fennoscandia. Prefers open uplands, moors mainly in Scotland, Scottish Islands, NW England and NW Ireland. More widespread, coastal and lowland habitats in winter.

♂

♀ imm.

Hobby *Falco subbuteo* 30–36cm

Kestrel-sized, but shorter-tailed, almost **Swift**-like shape and long, pointed wings. Fast aerial hunter like **Merlin**, 'hawks' for insects on the wing. Adult slate grey above, white below with heavy black streaks from throat to rusty brick-red thighs and vent 'trousers'. Head shows dark grey hood and bold, narrow dark moustachial stripe through white face. In flight, dark grey above, appears dark below at distance with heavy body and underwing mottling. White neck-patch and red vent notable at close range. Juvenile browner above, cream below, often no red 'trousers'.

Where to see: Scarce summer visitor, Apr–Sep winters in Africa. Breeds mainly S and E England, expanding N, some now in Scottish Highlands. Woodland nester, hunts in open countryside, farmland, wetlands and heathlands.

Birds of prey

Peregrine *Falco peregrinus* 36–48cm

Large, compact, powerful falcon with relatively short tail and broad-based wings. Adult slate grey above, white below with uniform grey barring. Dark grey crown and moustachial stripe contrast with white cheek, throat and upper breast, the latter with limited dark spots. Legs and eye-ring yellow. In flight, dark grey above with paler rump, paler below with fine dark barring throughout. Head pattern notable at distance. Juvenile similar, browner above with pale feather fringes, buff-cream with dark streaks below, pale tail-tip visible in flight.

Where to see: Scarce, widespread resident and winter visitor from N Europe. Favours upland moorland, rocky coasts, urban buildings to breed, mainly in N and W Britain, also varied open wetlands, marshes and urban areas in winter.

imm.

Barn Owl *Tyto alba* 33–35cm

Unmistakable medium-sized 'white' owl. Upperparts, nape and crown warm orange-brown-buff with intricate grey-black-white flecks throughout. Large head, face a heart-shaped white disc, eyes small, black and beady. Bill small and pale, mostly hidden under large white feathered 'V'. Underparts and underwing white. Ghostly appearance in half light of dusk, mainly hunts at night. In flight, warmer-orange buff to upperparts and outer wing, the latter also boldly barred, otherwise clean white. Hunts low over open grassland and along fringe habitats, hovers and dives on small mammal prey.

Where to see: Uncommon, widespread resident. Found in open countryside, grasslands, farmland, marshes and woodlands. Nests in tree holes, nest boxes and old buildings.

Tawny Owl *Strix aluco* 37–39cm

Large nocturnal owl with big, round head, stocky body, short wings and tail. Overall very brown, can be well camouflaged, roosting by day against a tree trunk or nestled in ivy. Adult mottled warm-brown, dark-brown and buff above, with white shoulder-patches and broad dark bars through wings. Underparts paler buff, with fine criss-crossed brown streaks. Flat-faced with large buff, brown-fringed disc and brown crown-stripe down face to bill. Eyes large, black, rimmed pink. Bill green-yellow. Call a sharp *ke-wick;* also a typical 'hoot'. Hunts by 'perch-and-pounce' attack. Grey, fluffy chicks leave nest at early age and climb into nearby branches.

Where to see: Fairly common, wide-spread resident, absent from Ireland. Nests in mature tree holes and also nest boxes. Found in deciduous woodlands, parks and large gardens.

Little Owl *Athene noctua* 21–23cm

Our smallest owl, **Starling**-sized, with compact profile, round body, short wings and tail. Dumpy profile can change to upright, leggy stance when alert. Brown above with white mottled patches, and white below with bold brown streaking. Facial plumage can give angry expression, with black rimmed, piercing yellow eyes, topped with bold white eyebrows above and white neck-band below.

Where to see: Scarce, declining (originally introduced) resident, mostly England, S Scotland, rare in Ireland. Often active by day as well as dusk. Nests in mature tree cavities and derelict buildings. Found in parkland, farmland and woodland edges. Frequently perches on branches, stone walls and farm machinery.

Long-eared Owl *Asio otus* 35–37cm

Large, upright nocturnal owl with round head and pronounced ear-tufts. Mottled brown, grey and buff, with dark criss-crossed brown streaks and speckling throughout. Brown bars through orange-brown wings and tail. Warm buff and cream below with dark streaks. Facial disc warm-buff, with a dark-fringed thick white 'V' between bright orange eyes. Two common postures adopted: tall pose with long features when roosting or threatened; more rotund, compact pose when relaxed. Lacks pale/white trailing edge to wing in flight.

Where to see: Scarce, widespread resident and winter visitor from Fennoscandia. Often roosts communally. Found in pine forests, sheltered conifer belts, thickets and woodland edges bordering open habitats, both inland and coastal.

Short-eared Owl *Asio flammeus*   33–40cm

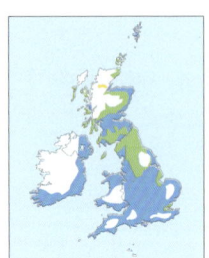

Large, often day-flying owl, with mottled straw-brown-buff plumage and very small ear-tufts. Upperparts brown with straw-buff patches, and warm brown heavily barred wings and tail. Pale buff below, with heavy dark brown streaks on breast, sparser on belly. Large head with pale facial disc and bright yellow eyes set in black eye-patches. In flight, like **Long-eared Owl** but pale bellied, without dark streaks, and light buff trailing edge to upperwing, not brown. Perched pose angled, not upright like Long-eared and **Tawny Owl**.

Where to see: Scarce, widespread resident and scarce winter visitor from NW Europe. Main stronghold uplands in NW England and Scotland, rare in Ireland. More common and widespread in winter. Nests on ground in open rough grassland, heaths, fens, marshes and moors. Favours rough coastal grasslands in winter.

Hoopoe *Upupa epops* 26–28cm

Unmistakable, with unique upright profile and exotic plumage. Large crest on small head, slender body, rounded wings and long, thin, decurved bill. Head and body coral pink. Wings, rump and tail black with bold white bars throughout; black and white particularly striking in flight when wide 'butterfly' wings extend fully. Crest long, often held flat, can raise into large coral-coloured fan with black tips. Feeds on the ground. Flight buoyant, rhythmic, low. Call a soft, hollow, far-reaching *hoop-hoop-hoop*.

Where to see: Rare, annual migrant from S Europe, Mar–Oct, winters in Africa. Mostly in spring, S and E coast England, more widespread during autumn. Bare, open ground, short coastal grassland and woodland edges.

Kingfisher *Alcedo atthis* 16–17cm

Striking coloration and small size enable easy identification. Small and compact, with short body, wings and tail, large head and weighty, dagger-shaped bill. Brilliant blue-aquamarine upperparts with lighter electric blue to rump and back. Mottled blue crown, blue facial stripe, orange-and-white ear-coverts and white throat. Underparts bright orange, short legs red. Bill dark, but lower mandible orange in female. Juvenile has duller plumage. Flies fast, straight and low, often just a flash of colour seen, dives into freshwater for small fish. Nests in riverbank holes.

♂

Where to see: Fairly scarce, widespread resident, absent in N Scotland. Freshwater habitat fringes, lakes, rivers and ponds, more brackish coastal habitats in winter.

♀

European Bee-eater *Merops apiaster* 27–29cm

Slender, exotic-looking bird. Unmistakable mix of rich chestnut orange and yellow upperparts, clean yellow throat with narrow black band and dazzling, uniform aquamarine underparts and tail. Bill slender and curved, face with jet-black eye-stripe and ear-coverts, tail long with longer central tail feathers. Juvenile similar but duller. Call a soft *prrut*. Nests communally, excavates nest-holes in sandy banks.

Where to see: Rare migrant from S Europe, Apr–Jul, mainly S and E coast England; winters in Africa. Very rare recent breeder in England.

Wryneck *Jynx torquilla* 16–17cm

Cryptically plumaged small woodpecker. Small bill, short legs, broad, long tail. Plumage 'bark-like' grey, brown, black, buff. Greyish above, brown stripe through face and neck, another along back. Tail grey, square-ended, with wide brown bars. Wings with dark criss-crosses and buffy spots. Fine dark barring to pale throat and belly. Perches horizontally, often on ground.

Where to see: Scarce annual passage migrant from NW Europe to S Europe, Africa, Apr–May, more frequently Aug–Oct. Majority coastal, S, E England, E Scotland. Gardens, verges, dry-stone walls and rocky open scrub.

Lesser Spotted Woodpecker *Dryobates minor* 14–15cm

Very small pied woodpecker, often elusive, hard to encounter. Black above with white bars across wings and back, no white shoulder-patches like **Great Spotted Woodpecker**. Underparts uniform off-white-buff, with some fine dark streaks to upper breast and flanks. Vent off-white, not red like Great Spotted. Male has red cap, female's black and white. Juvenile buff-brown below with more fine streaking throughout. Repetitive, loud *keek* call and distinctive drumming in spring, higher pitched and longer than Great Spotted. Rather fluttery flight, often for short distances at height within canopy. May associate with mixed woodland bird flocks in winter.

Where to see: Rare and declining resident in England and Wales. Found in lowlands deciduous woodlands, parklands, hedgerows and trees.

Woodpeckers

Great Spotted Woodpecker *Dendrocopos major*   23–26cm

Blackbird-sized woodpecker with black-and-white upperparts and uniform whitish underparts with scarlet undertail 'vent'. Back all black with broad white shoulder stripe and bold white spots on black wings. Head white with black cap and facial strip from bill across to nape. Male has red patch on nape. Bill rather stout, grey. Juvenile similar, with browner wing feathers and a red crown-patch. White shoulder-patch conspicuous in flight. Call a loud, abrupt *tchick, tchick,* spring territorial sound is a loud, rather hollow drumming of bill hammering tree trunk.

Where to see: Common throughout the British Isles, found in coniferous and deciduous woodlands, parklands, orchards and gardens. A frequent visitor to garden bird feeders.

♀

♂

♂

Green Woodpecker *Picus viridis* 31–33cm

Large, robust ground-feeding woodpecker with dagger-like bill. Mossy green above, with white spotting to blackish wings and striking yellow-green rump, obvious in flight. Large head has red crown, black mask around stark, pale eye, and black moustachial stripe, with red centre on male only. Cheek, throat and underparts pale cream, tinged green. Juvenile plainer-faced, no black face mask or moustachial stripe, duller green above with buffy spots and greyish, finely barred underparts. Awkward upright ground stance with stiff, short-legged hops. Can extend tongue into ant nests. Nests in tree holes. Call is loud, laughter-like yelping.

Where to see: Common resident throughout the British Isles. Found on open grasslands, parkland, heaths and garden lawns.

Ring-necked Parakeet *Psittacula krameri* 37–43cm

Exotic with slender body, round head, short wings and very long, spiky tail. Bright green above, more lime green below. Nape and tail with blue hues. Bill chunky, red, smoothly downcurved and sharply hooked. Adult male has narrow black and pink collar, curving neatly from chin to nape. Female has no face-ring or blue plumage hues. Juvenile like female but yellower and shorter-tailed. In flight distinctive, with long, thin tail and contrasting black-and-yellow underwing. Call a loud, unmistakable screech, often seen in flocks. Nests in adopted tree holes.

Where to see: Escapee, now established, locally common and expanding rapidly from original populations (mainly in London). Mainly urban and suburban towns, cities, parks, gardens, woodlands and orchards.

Red-backed Shrike *Lanius collurio* 16–18cm

Almost **Starling**-sized, with distinct shrike profile. Adult male shows chestnut-brown back and wings, grey rump and black tail with white edging. Head ash grey with bold black face mask, white throat and hooked black bill. Plain creamy-white below, tinted pink. Female duller, rufous-backed, with brown wings and tail and greyish plain rump. Greyish crown, plainer face, with brown ear-coverts, grey bill and dusky chin. White below with fine brown barred flanks. First-winter finely barred and scaly brown on back, crown and flanks.

♀

Where to see: Scarce annual passage migrant, most in autumn, from N Europe and Scandinavia, winters in Africa. Formerly widespread breeder, now very rare. Mostly S and E coasts, in open heath habitats, hedgerows, bushes and plantation edges.

♂

Great Grey Shrike *Lanius excubitor* 21–26cm

Large, pale grey and black shrike, upright stance, stocky body, short wings, long tail. Adult very pale grey throughout. Round head shows solid black eye-mask and white chin. Bill dark grey, hooked. Wings black with white wing-patch. Tail black with white outer tail feathers. In flight, pale grey above and on rump, contrasting black tail and wing, obvious white wing-bar. Juvenile fine dark barring throughout. Like all shrikes, preys on small mammals, birds, insects, often impales food for later consumption (larder). Perches atop vegetation, sometimes hovers.

Where to see: Scarce annual passage migrant and winter visitor from N Europe, Fennoscandia, winters in Africa. Found along coasts, most in autumn, more inland and widespread in winter. Forest edges, plantations, heaths and open ground with scrub.

Woodchat Shrike *Lanius senator* 18cm

Stocky, fairly short tailed shrike. Male black above with white shoulder, wing patch and rump. Chestnut cap, black face mask and black bill. Adult female grey-brown above, chestnut cap and nape, white lore patch within black eye mask. In flight, shows white rump and wing bar. First winter like first winter **Red-backed**

Shrike, but cold grey throughout with extensive barring including through white rump.

Where to see: Very scarce annual migrant from Europe, winters in Africa. Mostly spring, S and E coast. Similar open habitat to Red-backed Shrike.

Golden Oriole *Oriolus oriolus* 19–22cm

Strikingly colourful, **Blackbird**-sized but with longer wings and body. Adult male canary yellow, with black wings and lore patch and stout red bill. Adult female and first-winter dull yellow to olive-green above, brown-grey wings, and whitish below with fine dark streaks, yellow flanks and undertail. Grey lore patch and muted red bill.

Where to see: Rare migrant, past very rare breeder from S Europe, Apr–Oct; winters in Africa. Most in spring around coastal woodlands.

Jay *Garrulus glandarius*   34–35cm

Well-built colourful corvid with a shy, skittish nature. Upperparts pinkish-grey through hindcrown, nape and back. Large white rump, long all-black tail. Wings black with white wing-patch and unique dazzling blue-and-black chequered outer wing-panel. Head white with fine black streaking to crown, pale eye and broad black sub-moustachial stripe. Underparts dusky pink with white undertail-coverts and vent. Bill strong and grey, legs sturdy and dull pink. Often seen in flight between woodland trees, showing white rump and broad dark wings with white wing-bar. Often caches acorns for later consumption, otherwise eats insects, small birds, eggs, seeds and nuts.

Where to see: Common, widespread resident, scarcer in far N. Found in deciduous woodland, parks and large gardens.

Magpie *Pica pica*   44–46cm

Very distinctive long-bodied black-and-white crow with an extremely long tail and bold, loud disposition. Plumage all black aside from a large, oval white shoulder-panel and well-defined large white belly-patch. Shows green-blue-purple iridescent gloss on dark wings and tail in good light. Rather ungainly, upright stance on sturdy black legs. In flight, quite exotic profile, with broad, diamond-shaped long tail and extensive white outer wing wedge and white inner shoulder-bar. Call abrupt, loud, repetitive *chaker-chaker-chaker* and *cha*.

Where to see: Common and widespread throughout UK and Ireland, absent from NW Scotland and northern islands. Found mainly in woodland, parks, gardens, open country, farmland, rural and urban habitats.

Chough *Pyrrhocorax pyrrhocorax*   39–40cm

Stunning corvid with glossy black plumage, a relatively small head, long, red, downcurved bill and sturdy red legs. In flight, long, broad wings are square-ended and deeply fingered, often held swept back. Flight pattern often buoyant, with aerobatic half dives and twists. Often in pairs or small flocks. Call a characteristic *chwee-ou*, not unlike **Jackdaw** but higher pitched, more insistent.

Where to see: Scarce, local resident of coastal habitats restricted to W coast UK, mostly W Wales, N to SW Ireland, SW Scotland and the Isle of Man. Small populations re-established in Cornwall, and re-introduced to E Kent. Nests in cliff crevices, old disused coastal buildings. Found along cliffs, rocky shores, coastal grassland and rugged islands.

Jackdaw *Coloeus monedula* 33–34cm

Smallest British crow, rather stocky, round-headed and bull-necked. Adults dark satin grey-black throughout, with a black cap and distinct silver-grey shawl around nape and neck. Eye strikingly pale, whitish. Bill stout and dark, nostrils covered in dark horizontal feathering. Juvenile duller dark brown-grey, darker eye and less defined grey shawl. In flight, all show rounded wings and grey-black underwing. Agile, acrobatic flight of soars and flaps, often in large pre-roosting flocks with **Rooks**. Forms large flocks in winter and roosts communally. Call a distinctive *chack*.

Where to see: Common, widespread resident, scarce in NW Scotland. Found in farmland, parks, coastal cliffs, suburban and urban habitats. Nests in tree holes, rabbit holes, cliff crevices, old buildings and chimneys.

Rook *Corvus frugilegus* 44–46cm

Large black crow with small head, steep forehead and rather long, tapering bill. Adult shows distinct bare pale grey skin around broad bill base, reaching dark eye. Plumage has satin purple-blue sheen in good light, feathers rather dishevelled and loose around upper legs, giving square-bellied look. Juvenile like **Carrion Crow**, with no bare grey skin at bill base, but more pointy, tapered bill, feathered nostril 'bulge', and skinnier neck and head profile. Call a croaky, flat *kaah* or *graah*, often repeated. Aerobatic and sickle-winged in flight with long, rounded tail. Gregarious, vocal crow, breeds in treetop rookeries, feeding and roosting flocks common.

Where to see: Common resident throughout the British Isles. Found on farmland, open countryside and in suburban areas.

Carrion Crow *Corvus corone* 45–47cm

All black, size comparable to **Rook**, smaller than **Raven**, with chunky black bill, thick neck, flat, feathered nostrils and smooth, low forehead and crown. Bill stout, deep, with slightly curved upper mandible, not thin and pointed like Rook or huge and prominently bristled like Raven. Often vigilant and shy, may form large flocks around food sources. In flight, tail quite short, square-ended, unlike round-tailed Rook or wedge-shaped Raven. Call and nesting sites like **Hooded Crow**.

Where to see: Common and widespread throughout British Isles, largely replaced by Hooded Crow in Scotland and most of Ireland. Widespread throughout countryside, woodland edges, farmland, coasts, urban and suburban areas.

Hooded Crow *Corvus cornix* 45–47cm

Distinguished from other corvids by black and grey plumage, otherwise like **Carrion Crow**. Black head, breast, wings and tail, contrasting with pale uniform grey (almost mauve grey in some light) throughout. Bill stout, deep, with slightly curved upper mandible. In flight, tail quite short, square-ended. Calls vary, but include deep conk and croaky, repeated kraa notes. Often vigilant, shy, less so in urban areas. Large stick nest deep within trees, on cliffs, quarries, also on ground in isolated moorlands. Some hybridisation occurs where range overlaps with Carrion Crow.

Where to see: Northern species, widespread and common resident, largely replacing Carrion Crow throughout Scotland, Scottish Islands and most of Ireland. Scarce winter visitor from Fennoscandia E coast England. Open countryside, woodland edges, farmland, moorland, coasts and mountains, also urban and suburban areas.

Raven *Corvus corax* 60–68cm

Bulky, powerful corvid, bigger than **Common Buzzard**, characteristically shy. Plumage all black, bill huge, black and arched, dominating head. Beard-like throat feathers give 'front-heavy', bulbous appearance. Forehead flat, hindcrown rounded, feathers over nostrils bristly, flat against bill. In flight, long, diamond-shaped tail and long, broad wings with narrow-fingered tips. Very acrobatic, often soars at height then stoops with full roll tumbles and outstretched leg descents. Usually seen in pairs or small feeding groups in winter. Call varied, includes deep, low, croaky rattles, *prrruk* notes and repeated *krark-krark-krark*.

Where to see: Locally common, wide-spread resident. Scarcer in E England though range expanding. Found on uplands, moors, forests, rugged coasts, cliffs, farmland and buildings.

Waxwing *Bombycilla garrulus* 18–21cm

Uniquely patterned, compact, perched and flight silhouette profile similar to **Starling** with short legs and bill. Adult male, dusky-pink-grey throughout with broad crest, orange cheek and forehead, and black eye-mask and bib. Wings black with bright yellow-white 'U'-shaped feather-tip fringes, white wing-bar and cherry red waxy tipped secondary feathers. Tail-tip banded yellow, undertail rusty-red. Female similar but black bib diffused. Juvenile similar but white-yellow stripe on wing, no 'U' curve around individual tips, and red tips reduced. Feed in flocks, gorging on apples, berries, often confiding. Call a trilling *sirrrr*.

♂

Where to see: Scarce, irregular winter visitor from Fennoscandia, Nov–Mar. Mostly Scotland, N England and along E coast UK. Hundreds in irruption years when food shortages in Scandinavia force birds south.

1st win.

Coal Tit *Periparus ater* 11–12cm

Similar in size to **Blue Tit**, but with duller grey-black and buff plumage, larger head and shorter tail. Head and bib black, with large rectangular white face-patch through cheek and distinctive white stripe from hindcrown to nape. Olive-grey above with two distinct narrow wing-bars. Whitish underparts and dusky-pink buff through flanks. Juvenile duller, white face-patches and wing-bars with greenish-yellow tinge, bib smaller. Call a sharp, high *pea-shew, tsee* or thin *see*.

Where to see: Common, widespread resident throughout UK. Winter visitor, with variable influxes from N Europe, mainly SE England. Mainly conifer woodlands but also oak, birch woods, mixed forest, parks, gardens and suburban habitats. Readily comes to feeders, nests in nest boxes as well as tree holes.

Crested Tit *Lophophanes cristatus*   10.5–12cm

Small buff-brown forest tit with unique pointed crest. Plain buff-brown above, white below, with light peachy-buff flanks. White head bordered by narrow black 'bridle' band extending from nape around collar and encompassing bib. Small dark red eye, and black eye-stripe extending behind eye and curving down, around cheek. Spiky crest flecked black and white, cheek tinged grey. Call often reveals presence and includes high-pitched, excitable, purring trill, *trrrrr* and *bbrreee* repeated. Elusive, though readily visits feeders, sometimes seen in mixed winter tit flocks. Often feeds high in canopy.

Where to see: Scarce, very local resident of N Scottish pine forests, both old Caledonian and Scots Pine plantations. Can be fairly common within restricted range.

Marsh Tit *Poecile palustris* 11.5–12cm

Small, cool-brown above, neat, glossy black cap, small, square black bib. Face white, underparts off-white, peachy buff wash to flanks. Shows small pale patch at bill base. Call upbeat, high-pitched single *phicay* or *pitchew*. Song simple, rattling single *chip* or double noted *whit-a-whit-a-whit-a* repeated. Readily visits feeders, often in mixed winter tit flocks.

Where to see: Uncommon, declining resident, mainly England and Wales, absent from Ireland. Deciduous woodland, parks, large, wooded gardens, wooded farmland, damp wetland thickets and good understorey cover.

Willow Tit *Poecile montanus* 12–13cm

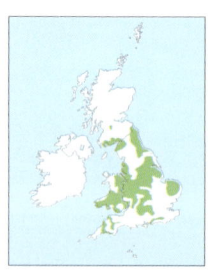

Like **Marsh Tit**, but stockier, thicker necked. Usually warmer buff-brown back, greyer wings, pale wing panel seen on closed wing. Black cap more matt than glossy, black bib larger, diffuse at edges, not clean cut, and square like Marsh Tit. Whitish below, peachy-buff to warm-orange flanks. Call squeaky *tsi-tsi-tsi* and nasal, buzzing *tzuah-tzuah-tzuah*, never *pitchew* like Marsh Tit.

Where to see: Rare, declining resident, mostly N and C England, Wales, S and C Scotland. Absent from Ireland. Woodlands, conifer and mixed, especially birch, alder and willow, damp wetland carr, dense thickets, hedgerows and parks.

Blue Tit *Cyanistes caeruleus* 11–12cm

Familiar, bright yellow-and-blue tit. Small, compact, with rounded body, short wings and tail. Back blue-green, tail blue, wings blue with small white wing-bar. Head white with electric blue cap, narrow black eye-stripe and face collar, the latter from nape to bib. Lemon yellow below, with narrow black belly stripe. Bill and legs grey. Juvenile duller with yellow wash. Nests in tree holes, readily uses nest boxes. Commonest bird feeder visitor. Very acrobatic and agile when perched and feeding. Call a thin *see-see'see*, alarmed *ker-rr-rr-rr*, song two sharp notes, then stuttered trill *sih-sih-surrrrr*.

Where to see: Common, widespread resident, winter visitor from N Europe. Frequents parks, gardens, woodlands, bird feeders, suburban and urban habitats.

juv.

Great Tit *Parus major* 14–15cm

Striking, beautifully bold tit, similar to **Blue Tit** but much larger. Back green, tail and wings blue-grey with obvious white wing-bar. Head glossy black with large white cheek-patch. Underparts lemon yellow throughout, with broad black stripe running from throat to underbelly (stripe narrower and peters out on belly in slightly duller female). Juvenile duller with yellow wash throughout. Call includes a single-note *tink*, song loud, sharp repeated *teacher-teacher*.

Where to see: Common, widespread resident, winter visitor from N Europe.

Found in parks, gardens, woodlands, at bird feeders, urban and suburban habitats.

Bearded Tit *Panurus biarmicus* 12–13cm

A shy, reedbed-loving bird with small, rounded body and very long tail. Adult sandy-brown (reed colour) above and below, with short black, sandy brown-and-white streaked wings. Adult male has blue-grey head and broad, tapering, vertical black moustache contrasting with white throat. Long tail sandy-brown, slender and graduated, undertail-coverts black. Small yolk yellow bill, pale straw eye and black legs. Female has pale sandy-buff head with two dark crown-stripes, white throat and undertail-coverts. Juvenile like female, but with black back stripe, outer tail stripes and lores. Call distinctive, a lively *ping*.

Where to see: Scarce resident, scattered, restricted coastal reedbed locations and other wetlands in winter. Irruptive autumn influxes from the continent can increase populations in winter.

♂

♀

Long-tailed Tit *Aegithalos caudatus*   13-15cm

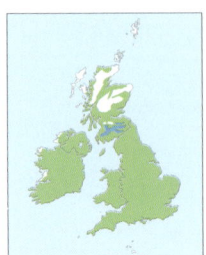

Unmistakable tiny bird with a very long, slender tail that comprises over half its total length. Upperparts mainly black, with black back, rose-pink shoulder, black wings with a pale wing-panel, and black tail with white outer-tail feathers. Head white with a broad black stripe either side of white crown, from bill across face, joining black nape. Underparts white with a rose-pink flush to flanks and underbelly. Small bill and legs black. Eye-ring dark pink. Fast and aerobatic, often moving through trees in small parties. Call a loud, thin *zee-zee-zee*, soft *pit* or high trill.

Where to see: Common, widespread resident, found in parks, gardens, mixed woodlands and scrub habitats; readily visits garden bird feeders.

Woodlark *Lullula arborea* 14–15cm

Small, short-tailed. Streaky brown above, with buff-fringed feathers. Short, dark streaked crown-crest, bold, pale stripe above eye, rufous cheek. Buff white below, dark breast streaks. Black-and-white pattern on closed wing edge, becomes white-edged black forewing patch in flight. Tail dark with white corners. Call fluty *tew-leet*, song, sweet, descending, includes *lu-lu-lu-lu*, often during high display flight. Readily perches in trees.

Where to see: Scarce resident, local, partial migrant. More widespread in winter, some move to SW England and continent. Mainly S, C England, Feb–Sep. Heathland, sandy soils, farmland and conifer plantation edges.

Skylark *Alauda arvensis* 18–19cm

Small, brown, streaky lark, larger and longer-tailed than **Woodlark**, much larger than **Meadow Pipit**. Upperparts brown, with pale brown edges to brown feathers. Head pale with dark streaky crown-crest. Face shows pale eyebrow and eye-ring and brown-buff cheek-patch. Upper breast heavily streaked dark brown. Belly buff white. In flight, white trailing edge to wings and broad white outer tail feathers obvious. Readily hovers. Call a chirrup, song loud, lengthy, with varied melodious trills, often performed in flight and high up.

Where to see: Locally common, declining resident, partial winter migrant to N Europe, Sep–Mar. Many in south UK don't migrate. Prefers arable and cultivated open farmland, grasslands, heaths and marshes.

Shore Lark *Eremophila alpestris*   16–19cm

Smaller and shorter-legged than **Skylark** with slimmer profile. Adult tawny-brown, subtly streaked above, clean white below. Breeding male shows bold yellow and black head markings and two protruding black feathers each side of black cap, which can be raised to form 'horns' ('Horned Lark'). Face bright yellow, with wide black mask-stripe from bill to eye and down through cheek. In winter, sexes similar, but head markings duller and horns absent. In flight, shows white underwing and black tail. Low ground-feeding profile, hunched, often inconspicuous but can be approachable.

Where to see: Rare, very local winter visitor along E coast Scotland and England from N Europe, Oct–Mar. Frequents shingle beaches, marshes and sand dunes, usually in small flocks.

Sand Martin *Riparia riparia* 12–13cm

Small, compact pale brown-and-white martin, with very shallowly forked tail and pointed, triangular-shaped wings. Adult brown, underparts white from bill to undertail broken by broad brown upper breast-band, visible at distance. Very small, dark bill and short, dark legs. Juvenile similar, but underparts more buffy, breast-band diffused and upperparts scaly with pale feather fringes. Nests colonially in holes in vertical sandy banks, cliffs and quarries. Readily perches on wires, often with **Swallows**. Hunts on the wing, low over open countryside, wetlands and reedbeds. May migrate and roost in large flocks.

Where to see: Locally common, widespread summer visitor and passage migrant, Mar–Oct, winters in Africa.

Swallow *Hirundo rustica* 17–19cm

Slender with elongated profile, pointed wings and deeply forked tail with tail-streamers (longer and wire-like in male). Adult black above, with iridescent blue sheen. Forehead and throat deep red, upper breast-band broad and dark. White below, often tinged warm buff. Small white 'mirrors' visible on tail when fanned. Juvenile duskier, throat more orange-buff, tail-streamers short. Tiny, dark bill, very short, dark legs. Agile, fluid flight, often very low, hunting insects on the wing. Readily perches on wires. Nest an open mud-straw bowl on internal old building rafters, in stables and barns. May migrate in large flocks.

Where to see: Common, widespread summer visitor and passage migrant, Mar–Oct; winters in Africa. Farmland, open countryside, hedgerow edges, wetlands, rural and suburban habitats.

House Martin *Delichon urbicum* 12–13cm

Unmistakable black-and-white martin, compact, with a forked tail, short wings and white rump. Adult upperparts and 'hood' almost black with a faint blue gloss, wings dark brown. Rump-patch square and white. Underparts pure white from chin to tail. In flight, underwing all dark, contrasting with pure white underparts and black tail fork. Legs short, feathered white. Juvenile similar but duller. Hunts insects on the wing, often at height. Nests colonially, nest is a mud cup often built under building eaves, also uses species-specific nest boxes. Readily perches on wires.

Where to see: Locally common, widespread summer visitor, Mar–Oct, winters in Africa. Scarcer in Scotland. Favours suburban and semi-rural environments.

Red-rumped Swallow *Cecropis daurica* 16–17cm

Similar to **Swallow** but orange-white rump. Adult has blackish wings and blue-black sheen to back and cap. Nape rusty-red, cheeks, throat and underparts cream, finely streaked. In flight, undertail-coverts dark from below, not pale like Swallow, and no white mirrors to tail. From above, dark tail and tail base contrast with orange-white rump and dark upperparts. Long tail streamers, may appear as one spike. Juvenile similar, with scaly pale fringes above, duller rump.

Where to see: Very scarce annual visitor from S Europe, Apr–Sep; winters in Africa. Mainly S and E coast.

Cetti's Warbler *Cettia cetti* 13–14cm

Small, stocky, with short, rounded wings, relatively long, broad-ended tail, often held cocked. Warm rufous-brown above, greyish cheek, whitish line above eye. Greyish below, whiter on throat and belly. Skulking nature, presence often only revealed by song, a loud, explosive series of liquid, fast notes including *pwit pity-chewit, chewit, chewit* from dense cover. Often only glimpsed in flight, low between bushes.

Where to see: Uncommon, local resident, scattered distribution through England and Wales, most S, E England. Coastal marshes, reedbeds, ditches, dense scrub and vegetated riversides.

Wood Warbler *Phylloscopus sibilatrix*   11–12.5cm

Larger than **Willow Warbler**, longer wings, shorter tail, starker, brighter contrasting plumage. Upper-parts bright olive green, pale green-yellow edges to grey wing feathers. Pale, beady-eyed face, lemon yellow, broad eye-stripe, cheek-patch and throat. Underparts clean white, some yellow to flanks and upper breast. Song often heard from high woodland, open perches, a fast, sweet, accelerating jangling trill.

Where to see: Scarce, local summer visitor from Africa, Apr–Sep. Scattered distribution, mostly W England, Wales, Scotland. Upland mature oak woodlands with limited understorey, often at height.

Hume's Warbler *Phylloscopus humei*   9–10.5cm

Like **Yellow-browed Warbler**, previously regarded as a subspecies. Overall duller, with pale grey-green tones. Face grey-buff, washed out, with less contrasting pale yellow-whitish-buff supercilium. Mid wing-bar paler, yellow-buff, upper wing-bar reduced to small spot or absent. Bill and legs darker. Call essential for identification, very different from Yellow-browed Warbler, dull, plaintive sweet *che-wee* or down-slurred harsher *swee-oo*.

Where to see: Rare migrant from Asia. Mostly late autumn, Oct–Nov, like Pallas's Warbler, occasionally overwinter. Most records E and S coast, scrub and woodland edges.

Warblers

Yellow-browed Warbler *Phylloscopus inornatus* 9–10.5cm

Small, well-marked warbler with distinct yellow wing bar and supercilium. Greenish above, yellow fringed, grey-black wing feathers, yellow mid wing-bar, short yellow upper wing-bar. Tertials grey, white-tipped. Head with bold yellow supercilium. Eye-stripe and plain crown dark, ear-coverts mottled green and cream. Underparts whitish. Often vocal, call a sharp, thin, rising tchu-wee or tsweet, repeated. Fast, agile, flits and flutters, feeding within leafy tree canopies.

Where to see: Scarce vagrant from Asia, Sept–Nov, occasionally overwinters. Coastal scrub, woodland edges. Most S and E coast Britain from Isles of Scilly to Shetland.

Pallas's Warbler *Phylloscopus proregulus* 9–9.5cm

Similar to **Yellow-browed Warbler**, but bright, green-olive above, clean white below, with lemon yellow-white rump-patch, most obvious in flight. Head patterning like Yellow-browed Warbler but with bright yellow central crown-stripe, brighter yellow supercilium, with darker almost black eye-stripe and darker olive crown. Active, agile, often feeds among leafy tree canopies. Not as vocal as Yellow-browed Warbler, often a short, soft, rising *chuee*.

Where to see: Rare late autumn migrant from Asia, most Oct–Nov, occasionally overwinter. Coastal scrub and woodland edges, E and S coast from Isles of Scilly to Shetland.

Willow Warbler *Phylloscopus trochilus*   11–12.5cm

Small, slim greenish-brown warbler, cool brown-grey to olive-grey above and beige-white below, whitest on belly. Similar to **Chiffchaff** but longer winged, giving more elongated profile, and more striking head pattern, with stronger supercilium, as pronounced behind as in front of eye, and a more defined dark eye-stripe. Although variable, often has strong yellow wash to face, supercilium and underpart and usually shows pale pink-brown legs. Juvenile brighter, stronger yellow tones to underparts. Uniquely moults twice a year before each long migration. Call a clear *hooeet*, song fluid, sweet, with rapidly descending notes.

Where to see: Common, widespread summer migrant, Apr–Oct, winters in Africa. Prefers scrub and woodland, open areas with scattered trees, coniferous plantations and woodland fringe habitats. Favours birch woods in Scotland.

Chiffchaff *Phylloscopus collybita*   10–12cm

Similar to **Willow Warbler**, can be challenging to separate. Has shorter wings and more rounded profile, rounder head, dark, almost black legs and darker bill. Also regularly dips tail. Both dark eye and pale eye-crescent more obvious as set in weaker head pattern than Willow Warbler. Upperparts olive green-brown, with brown wings showing greenish paler fringes. Underparts beige-white with dull dusky-yellow wash to flanks. Juvenile browner, with more yellow to underparts. Call a thin, soft *hweet* repeated, song a steady, repeated *chif-chaf-chif-chaf*.

Where to see: Common, widespread summer and passage migrant, Mar–Oct, winters in Africa and Europe. Some birds from N Europe winter in Britain, some British birds now stay all year. Frequents woodlands, scrub, open habitats with scattered trees, parks and gardens.

Sedge Warbler *Acrocephalus schoenobaenus* 11.5–13cm

Small, streaky brown warbler with rounded tail and rather flat forehead. Pale brown above with dusky streaked back, and dark brown wings with pale cream feather fringes. Head shows bold, long cream supercilium contrasting with finely streaked dark crown and dark eye-stripe. Underparts plain creamy-white, whitest on throat, warm buff on flanks. Unmarked warm brown rump obvious in flight. Call a sharp, rolling *errr*, or *tuk*, song more varied than **Reed Warbler**, with a long, excitable mix of trills, whistles, squeaks and grating, sometimes also mimics.

Where to see: Locally common, widespread summer migrant, Apr–Oct, winters in West Africa. A wetland fringe species, from marshes to riversides, with suitable bramble, reedbed and thick vegetation.

Reed Warbler *Acrocephalus scirpaceus* 12.5–14cm

Very similar to very scarce **Marsh Warbler**, can be challenging to separate on subtle colour differences and structure. Plain, slender, unmarked brown warbler, with warm brown upperparts and whitish-buff underparts. Head shows rather flat forehead, an indistinct pale supercilium which stops at eye, and a longish fine bill. Throat white, longish tail and rump warm brown. Juveniles warmer toned through-out. Call often a rolling *churr*, song a long, repetitive churring, whistling and grating, *kerr-kerr-kerr-chiri chiri chiri*, like **Sedge Warbler** but more evenly paced and repetitive.

Where to see: Locally common, fairly widespread summer migrant, Apr–Oct, winters in Africa. Scarcer in Scotland and Ireland. Favours reedbeds, also found in fenland, ditches, marshes, riversides, other wetland fringe habitats.

Marsh Warbler *Acrocephalus palustris* 12.5–14cm

Uniform olive-brown through-out upperparts, lacking warmer toned, contrasting rump of **Reed Warbler**. Slightly longer, pointier wings, more obvious pale tips to primaries, paler tertial edges. Bill marginally shorter, broader base than Reed, legs paler straw yellow-pink. In autumn, Marsh can look paler, beige-brown above, creamy flanks. Reed shows warmer tones above on rump, richer buff below. Call a *chack* rather than a *churr*, song sweet, lively, lacking harshness, much more varied than repetitive Reed, incorporating much mimicry of bird species it encounters both in Europe and in Africa..

Where to see: Rare, local summer visitor and passage migrant, May–Oct, winters in Africa. Breeding now confined to a few locations in England, N Scotland. Most numerous along all E coast on passage. Favours dense, tall scrub, damp thickets and tall vegetation, not always near water.

Icterine Warbler *Hippolais icterina* 12–13.5cm

Large, robust warbler, strong, broad, pale orangey bill, chunky grey legs (see **Reed Warbler** and **Willow Warbler**). Grey-brown above with olive-green hues, pale buff-white below with variable but striking pale yellow tones, brightest in spring. May show flat forehead, can raise to peaked crown. Face open, unmarked, paler between eye and bill. Shows long primary projection, pale wing-panel from broad pale-edged wing feathers. Tricky to separate from rarer, more skulky Melodious Warbler (*H. polyglotta*, 12–13cm), but Melodious shows short primary projection, no pale wing-panel.

Where to see: Scarce, migrant from W Europe, most in autumn along E and S coast, from Shetland to Scilly Isles. Winters in Africa. Melodious rarer, mostly Shetland and S coast England, Ireland. Coastal scrub, trees and bramble, Aug–Oct.

Grasshopper Warbler   *Locustella naevia* 12.5–13.5cm

Slender, streaky brown, skulking warbler with flat forehead, short, round wings, and long, rounded, graduated tail. Olive-brown above with dark, soft streaking on back, wings, and rump. Crown finely streaked, supercilium buff. Underparts buff-white with diffused, dark streaking to breast, flanks, undertail-coverts. Streaky undertail feathers notably long. Call a sharp *tik*, song lengthy, high, reeling trill, likened to the call of a cricket.

Where to see: Scarce, widespread summer visitor, scarce passage migrant, Apr–Oct, winters in Africa. Dry and wet habitats, scrub, marsh fringes, rough grassland and reedbeds, more coastal on migration.

Savi's Warbler   *Locustella luscinioides*   13.5–15cm

Similar to **Reed Warbler**. Warm brown above, whitish-buff below, brown wash along flanks, whitish throat. Head with narrow pale buff stripe above eye. Wings shorter than Reed Warbler, tail broader-ended, rounder, graduated, with pale scaly feather edges to long undertail-coverts. Secretive, often only heard. Song like **Grasshopper Warbler**, insect-like prolonged buzz, but shorter, lower, and faster, blending into buzzy trill rather than discernible reel.

Where to see: Rare summer visitor, mostly S, E England, Apr–Aug, winters in Africa. Rare migrant, mainly S and E coast. Reedbeds and coastal scrub.

Blackcap *Sylvia atricapilla*   13.5–15cm

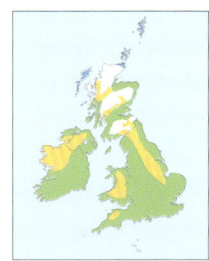

♀

Sturdy warbler with plain grey upperparts, browner grey on wings, and plain pale grey underparts. Adult male has black cap reaching eye and encompassing forehead. Adult female similar but with rufous-brown cap. Both show dark eye with small white lower eye-crescent. Juvenile has grey-brown cap, that wears away during first year to reveal adult coloration. All show silvery grey bill and legs. Call a hard *tak*, song beautiful, clear, rich and melodic warble that picks up pace.

Where to see: Common, widespread summer visitor, passage migrant from NW and C Europe, Mar–Nov, winters in Mediterranean and Africa. Increasing numbers overwinter. Found in deciduous and mixed woodland, also copses, thickets and parks. Garden bird feeders and apples in winter.

♂

Garden Warbler *Sylvia borin* 13–14.5cm

Small, stocky, washed-out grey-brown warbler. Cool grey-brown above, paler buff-white below. Head grey-brown, paler throat and collar. Face open, gentle expression, large black eye, faint, thin white eye-ring and eye-stripe, not always present (see **Barred Warbler**). Bill grey, sturdy, slightly chunkier than other small warblers. Legs dark grey. Juvenile similar but shows obvious pale fringes to tertial feathers. Call harsh, irritated *tack* or *churr*, song beautiful, melodic, very similar to **Blackcap** but slightly harsher, faster, with subdued, less clearly defined notes throughout.

Where to see: Locally common, widespread summer visitor and passage migrant from Fennoscandia, Apr–Sep, winters in Africa. Scarce, Ireland and N Scotland. Found in deciduous, mixed woodland, scrub, dense vegetation and parkland with good understorey. Not common in gardens.

Barred Warbler *Curruca nisoria* 15–16cm

juv.

Large, robust, slow-moving, pale grey-and-white warbler with rather harsh expression. First-winter birds scarce but annual, adults rare. Head and upperparts pale grey-brown, contrastingly buff-white narrow wing-bars, pale buff-white feather fringes to tertial and primary feathers. Creamy-buff below with variable fine grey bars to flanks and undertail-coverts giving scaly, scruffy appearance. Bill heavy with pale pink to base. Legs dark grey. Adult male very grey and white, white feather edges to grey upperparts and heavily grey barred white underparts. Female browner, less distinctly marked. Adults have pale eyes, juveniles dark.

Where to see: Scarce annual passage migrant, mostly E and S coast Britain, from Shetland to Scilly Isles. Summers in E Europe, winters in Africa. Loves scrub, dense vegetation, berry bushes and brambles.

Lesser Whitethroat *Curruca curruca*   11.5–13.5cm

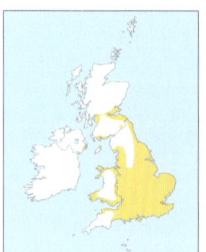

Small, compact warbler with rounded features and fairly short tail. Upperparts uniform brown-grey. Head storm grey with contrasting darker grey ear-coverts, looks almost hooded, white crescent to base of eye. Bill fine, dark, legs grey. Throat and foreneck white, underparts cream-white, warmer buff-pink wash to flanks in spring. Tail dark grey-brown with white outer tail feathers, obvious in flight. Autumn birds can look worn and stronger grey on upperparts. Juveniles often browner above, whiter below, obvious darker ear-covert patch. Call low *tuk* or *trrrr*, song repetitive, loud rattling and trilling, often from cover.

Where to see: Fairly common, widespread summer visitor and passage migrant, Apr–Oct, winters in East Africa. Larger range on passage. Prefers mature hedgerows, scrub, bramble and berry bushes.

Whitethroat *Curruca communis*   13–15cm

Slightly larger, longer-tailed than **Lesser Whitethroat**, more complex plumage, less rounded, more peaked crown. Adult male mid-brown above, rusty brown wing-panel from rusty-edged, dark-centred wing feathers. Head greyish through crown and ear-coverts, contrasting with white throat and pinkish-buff underparts. White broken eye-ring and pale pink bill base. Legs brown. Adult female and juvenile have brown not grey head, browner, duller on upperparts, show rufous orange-brown wing-panel. Call a *tac* or irritated *churr*, song rather jumbled, lively, though rhythmic, scratchy warble.

Where to see: Fairly common, widespread summer visitor and passage migrant, Apr–Oct, winters in Africa. More wide-spread than Lesser Whitethroat, extending through Ireland and N Scotland. Larger range on passage. Found in scrub, mature hedgerows, woodland edges, nettles and bramble.

Dartford Warbler *Curruca undata* 13–14cm

♀

Very small, short-winged warbler with long, slender tail, often held upright. Adult male dark grey-brown above, grey-blue through head, variable deep wine red through to rufous orange-brown wash through underparts, strongest around throat which also has small white spots at close range. Head with striking red eye-ring. Female and juvenile similar but browner, duller, less contrasting plumage, buffy underparts, paler throat, dull red eye-ring. Call a low buzzing, muffled *tchairr*, song not unlike **Whitethroat**, a lively, scratchy, and jumbled warble, though more metallic. Although secretive, often sings from tops of low gorse bushes.

Where to see: Scarce, very local resident of open, lowland heathland in S England. Also thinly distributed in East Midlands and Wales. Prefers gorse, heather and scattered low pines.

♂

Subalpine Warbler *Curruca* spp.   12–13cm

Moltoni's ♂

Eastern ♀

Eastern ♂

Western ♂

Currently three species, Western Subalpine Warbler (*C. iberiae*), Moltoni's Warbler (*C. subalpina*) and Eastern Subalpine Warbler (*C. cantillans*). Identification extremely difficult, often involving call recordings, tail-pattern differences, DNA analysis. Eastern commonest, a small, fine-featured warbler, adult male grey-blue above and on head, striking white sub-moustachial stripe, red eye-ring. Underparts whitish, variable brick red to pinkish-peach through throat, upper breast, flanks.

In flight, white outer tail feathers conspicuous. Adult female and first-winter duller, brown-grey upperparts, rufous-brown wings, subtle white moustache. Subtle white eye-ring around reddish eye. Pinkish tinge to throat, flanks. Pale pinkish-yellow legs.

Where to see: Very scarce annual migrant from SE Europe, Apr–Oct, mainly S/E coast Britain, winters in Africa. Most records in spring. Prefers coastal scrub, gorse and bramble.

Firecrest *Regulus ignicapilla* 9–10cm

Similar to **Goldcrest** but bolder head pattern. Upperparts moss green, shoulder bronzy-yellow, wings grey with pale feather fringes, contrasting white wing-bars and dark grey wing-patch. Head distinctive, with broad white stripe above black eye-stripe, grey ear-coverts, short black moustache and bright orange-centred black crown. Male's crown-stripe electric orange when raised and fanned. Underparts clean, whitish grey. Call and song like Goldcrest's, but song a simpler trill, with repeated *zi* notes increasing in volume and intensity. Juvenile duller, lacks crown-stripe.

Where to see: Scarce, local resident, mainly SE England and Wales. Scarce, more widespread coastal migrant, and winter visitor, Sep–Mar, from Europe. Prefers coniferous woodland and parkland to breed, more coastal scrub habitats on migration.

Goldcrest *Regulus regulus*   8.5–9.5cm

Tiny, rounded, short-tailed bird, rather like a *Phylloscopus* warbler but dumpier with striking plumage differences. Upperparts olive-green, wings dark grey, contrasting pale yellow fringes to feathers, prominent white wing-bars, dark wing-patch. Head pale grey, face plain, open, beady dark eye, short black moustache and narrow black crown with striking yellow centre. Hidden bright orange central crown-stripe on male, can be raised and fanned when alarmed or displaying. Underparts clean, grey-white to buff. Juvenile duller, lacks crown-stripe. Call repeated high-pitched *zsee*, song repeated thin trill and twittering, ending with jangling flourish. See also **Firecrest**.

Where to see: Common, widespread resident, migrant and winter visitor from N Europe, Sep–Mar. Prefers coniferous woodlands and plantations, parks, large gardens, and coastal scrub on migration.

Wren *Troglodytes troglodytes*   9–10.5cm

Tiny, rounded, energetic brown bird with short, regularly cocked tail. Warm brown above, fine dark bars throughout. Wings dark brown with fine buff-white bars. Head paler brown-buff, warm brown crown, narrow dark eye-stripe, broad buff-white stripe above eye. Fine dark bill slightly decurved. Underparts buff-white with narrow dark bars along flanks. Pale grey-pink legs. Flight rapid, straight and low between dense vegetation cover. Call loud *chac*, or grating, rattling *cherrr*, song elongated, loud, powerful trilling array of fast notes, often boldly from an exposed perch.

Where to see: Common and widespread resident throughout Britain in varied habitats: woodland, dense scrub, large and small gardens, parkland, hedgerows, open moorland and coastal cliffs. Some different island subspecies occur in N Scotland.

Nuthatch *Sitta europaea*   12–14.5cm

Great Tit-sized, with straight-backed body profile, short, thick neck, large head. Dark bill dagger-shaped, legs robust, tail short, square-ended and stiff. Male neat blue-grey above, including crown and tail. Face and throat white with bold, long black eye-stripe from bill base through eye and into nape. Underparts buff-orange, rusty-chestnut streaks to hindflanks and undertail-coverts, the latter also with white chevrons. Flight heavy, straight, tail shows black-and-white corners. Female and juvenile duller, buffy-peach below. Call short, loud, *doid* or *chwit*, song simple, clear *twee-twee-twee*. Agile, climbs jerkily up and down tree trunks and branches, often head-first.

Where to see: Fairly common, widespread resident, absent from Ireland, rare in N Scotland. Frequents woodland, parkland and gardens, comes readily to feeders.

Treecreeper *Certhia familiaris* 12.5–14cm

Small, delicate, brown-and-white woodland bird, long, fine decurved bill, short legs, long, stiff and pointed tail feathers used to aid tree climbing. Plumage cryptic against tree bark. Upperparts brown with pale buff-white mottling, wings darker brown with pale buff feather tips and fringes forming a stepped, barred pattern that appears as an obvious pale bar across wings in flight. Head has mottled brown-and-white crown, eye-stripe, uneven short white stripe above eye. Underparts clean white, tail and rump brown. Call thin, high-pitched *tsee*, song thin, trill and warble.

Where to see: Common, widespread resident of deciduous and coniferous woodlands, parks and large gardens throughout Britain and Ireland. Creeps up and around tree trunks and branches hunting insects, flies down to base of another tree to start heading up again.

Rose-coloured Starling *Pastor roseus*   19–22cm

Very like **Starling** in structure and behaviour, but plumage very different. Adult pale pink back, rump, underparts, contrasting with glossy black head, throat, upper breast, wings, tail. Crown has spiky black crest, often held down. Bill pink, stouter than Starling, legs brighter pink. Undertail-coverts blackish with white scaly fringes. Juvenile like Starling juvenile, but paler brown-grey throughout, wings darker, bill base yellow not black. Face plain, pale, no dark lores like Starling juvenile. In flight, contrasting pale greyish rump and dark wings. Often associates with Starlings.

Where to see: Very scarce annual migrant from E Europe, winters in Asia. Adults mostly May–Aug, juveniles Aug–Oct. Adults widespread, juveniles mostly S coast. Various habitats, like Starling.

juv.

Starling *Sturnus vulgaris* 19–22cm

sum.

Small, dark bird, short tail, has confident strutting walk. Adult summer black with iridescent purple-green sheen, pale buff wing feather edges, small buff-white 'V'-shaped feather tips giving light spangled, spotty appearance to back, rump, hindflanks. Bill yellow. Winter adult similar but bill black, and buff-white 'spangling' spots more extensive throughout all plumage. Juvenile plain pale brown-grey, paler feather fringes throughout. First-winter resembles winter adult but with grey-brown head and neck. All show pink-brown legs. Triangular-winged in flight. Forms huge aerial pre-roosting 'murmuration' flocks in winter. Call a *tcheerr*, song lengthy warbling mix of mimicry, whistling and rattling.

Where to see: Common, widespread resident and winter visitor from Europe Sep–Apr. Frequents many habitats: grasslands, parks, farmland, woodlands, coasts, rural and urban areas and gardens.

juv.

win.

Ring Ouzel *Turdus torquatus*   24–27cm

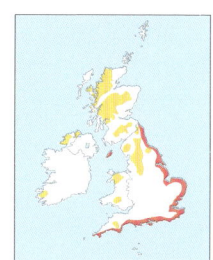

Large, dark thrush, like **Blackbird** but slender-bodied, with longer wings and tail. Male black with broad white breast-band, wings with silver-grey feather fringes, creating a pale panel at distance. Bill yellow, black tip. In non-breeding plumage has fine silver-grey feather fringes throughout, especially to underparts, giving scaly appearance, breast-band duller. Female like non-breeding male but browner throughout.

Juvenile brown, scaly, with white barring and pale spots to underparts. All show paler wings than body in flight. Call a hard *tac*, song similar to that of **Mistle Thrush**, strong, fluty, with pauses.

Where to see: Scarce, local summer visitor Mar–Nov, uncommon passage migrant from Fennoscandia, winters in Europe and N Africa. Breeds in uplands, more widespread and coastal on migration.

Blackbird *Turdus merula*   23.5–29cm

♀

Large, round-headed, long-tailed thrush. Male matt black with yolk yellow bill and yellow eye-ring. Female mid-brown with paler brown throat and upper breast often with subtly darker mottling and streaks. Bill variable brown-yellow. First-winter male dark bill and brown wings. Juvenile rich brown, extensive brown mottling and spotting to paler underparts. All show brown legs. Call loud, abrupt *chak* or alarmed *chink*, song rich, melodious warble, long and fluty, most vocal during dawn chorus.

Where to see: Very common, widespread resident, passage migrant and winter visitor from N Europe, Oct–Mar. Often seen on lawns, tail raised, head cocked, listening for worms and invertebrates. Prefers gardens, parkland, farmland margins, woods and hedgerows, rural and urban areas.

♂

Fieldfare *Turdus pilaris* 22–27cm

Large, robust, colourful thrush with a long tail. Head, nape, upper back and rump pale grey, lower back and wings chestnut-brown. Face shows indistinct whitish eyebrow, smudgy black lores and yellow, black-tipped bill. Throat and upper breast buff-cream to orange, with heavy black streaking and spotting that extends along flanks. Underbelly white, tail black. In flight, large grey rump, black tail and white belly and underwings obvious.

Call distinctive, loud *shak-shak-shak*, often heard in flight.

Where to see: Common, widespread and familiar winter visitor and passage migrant, from NE Europe and Fennoscandia, Oct–Apr. Also rare northern breeder. Often seen in large, noisy, roaming winter flocks. Frequents fields, orchards, large gardens, parks, farmland, hedgerows and woodland. Feeds on berries and invertebrates.

Redwing *Turdus iliacus* 19–23cm

Small, boldly marked, short-tailed thrush. Adult uniform mid-brown above, head with obvious pale cream eyebrow and sub-moustachial stripe, brown cheek, dark brown lores and yellow-based black bill. Underparts white, with heavy brown streaking from throat to underbelly and conspicuous, long rusty-red flank-patch. Legs pink. Underwing dusky cream with striking dark red-brown inner wing. First-winter birds have some pale buff tips to brown wing feathers. Call a thin *seep*, often heard at night during autumn migration.

Where to see: Common, widespread winter visitor from Iceland, Fennoscandia, NE Europe, Sep–Apr. Also rare N breeder. Often in fields, parks, gardens, orchards, woodland edges and hedgerows. Frequently found in mixed winter flocks with **Fieldfare**, feeding on open ground or in berry bushes.

Song Thrush *Turdus philomelos* 20–22cm

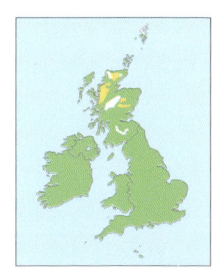

Small, spotty, familiar garden thrush with short tail. Upperparts uniform brown. Face pale brown-buff with cream lores, weak cream eyebrow, mottled ear-coverts, brown bill. Underparts whitish, creamier through breast and flanks, heavy dark brown droplet-shaped spots throughout. Juvenile shows some pale buff tips to brown upperparts. All show apricot-orange inner wing in flight and pinkish-brown legs. Call a high-pitched, clipped *sip* or *tik*, song rich, loud, fluty, many phrases repeated in twos and fours. Secretive nature, often alone. Unlike other thrushes, snails are a favoured food choice – may be heard hitting their shells against stones.

Where to see: Locally common, widespread resident, migrant and winter visitor from NW Europe, Oct–Mar. Found in gardens, parks, hedgerows, woodlands and field margins. Prefers feeding under cover.

Mistle Thrush *Turdus viscivorus* 26–29cm

Robust, large thrush, bold, upright stance, small head, barrel chest, long tail and wings. Cold grey-brown above, paler buff fringes to wing feathers creating wing-panel, pale creamy-buff below with dark brown rounded and crescent spots throughout. Shoulder has smudgy brown patch, face pale, small dark ear-covert crescent. Bill dark, strong legs pink. Juvenile shows extensive pale tips and fringes to upperparts. Flight strong, undulating, tail square-ended, white tail corners and underwing-coverts notable. Call a distinctive, dry rattle, song loud, melodic, more mournful and harsher than **Blackbird**, often from treetops.

Where to see: Fairly common, widespread resident, passage migrant and winter visitor, Sep–Apr. Absent from some areas of N Scotland. Prefers open woodland and farmland, fields, large parks and gardens in winter.

Spotted Flycatcher *Muscicapa striata* 13.5–15cm

aut.

Small, slender, upright flycatcher, short legs, long wings, rather uniform plumage. Adult pale grey-brown above, paler grey feather edges to darker wings. Head plain, finely streaked crown, black beady eye, black bill. Underparts white, subtle, fine dusky streaks to throat, breast, flanks. Legs dark. First-winter and juvenile have buffy spots to upperparts, juvenile also scaly above and below. Agile, often snatches insects on the wing from lookout perches, returns to perch to eat. Perches in open, wings flicking, tail dipping. Call short, high *tzee*, song mix of thin, simple scratchy, squeaky notes and trills.

Where to see: Uncommon, declining, widespread summer visitor from South Africa, May–Sep. Passage migrant from Fennoscandia. Prefers woodland edges, rides, parks and gardens; wider variety of habitats on migration.

Robin *Erithacus rubecula* 12.5–14cm

Very familiar garden bird, with striking plumage. Profile shows small, rounded head, long straight legs and either rounded or slender body, depending on temperature. Upperparts warm brown with olive tones. Forehead, face, throat and breast strong red-orange with variable, partial grey border. Underparts white from belly to undertail-coverts, flanks peachy-buff. Eye black, conspicuous, bill fine, dark brown-black, legs pink-brown. Juvenile coppery brown, pale feather fringe scaling above and dark scaling to underparts, orange-red breast plumage absent. Call sharp *tik* or high, thin *seee*, song a long, rich, melodious warble with melancholic tones. Sings all year.

juv.

Where to see: Common, widespread resident, passage migrant and winter visitor from NW Europe, Sep–Mar. Found in woodland, parks, gardens, hedgerows, scrub and suburban habitats.

Bluethroat *Luscinia svecica*   13–14cm

Small chat, male has stunning spring plumage. Adult male grey-brown above, head with bold cream-buff supercilium, buff-white below. Throat and breast royal blue, usually with white or red central throat spot followed by black and brick red bands through lower breast. Adult female and first-winter less colourful, most show cream-buff throat and upper breast, black lateral throat-stripes, breast streaks and spots. Many may show some diffused blue-and-red bands only alongside black band on lower breast. All show distinctive tail pattern, orange-brown tail sides, wide black tail-band, obvious in flight. Call loud, sharp *shak*.

spr.

Where to see: Scarce passage migrant from Fennoscandia, mainly autumn, Aug–Oct, but also Apr–May, winters in Africa. Coastal from Shetland to Scilly Isles, frequents gardens, reedbeds and scrub.

1st win.

Nightingale *Luscinia megarhynchos* 15–16.5cm

Shy, secretive chat with plain plumage, long wings and tail, typical upright chat stance on long legs, tail often held cocked. Adult warm brown above, striking rufous-brown rump and tail. Underparts buff-white, palest on throat and undertail vent. Face plain, beady dark eye, pale eye-ring. Bill dark, some yellowish-pink to base, legs long and brown-pink. Juvenile similar but fine dark feather edge scaling throughout. Often skulks in low, dense scrub cover, feeds and nests on ground. Known for beautiful, melodious, liquid song in spring. Call or alarm a low, grating *kerrr*.

Where to see: Rare, very local summer visitor to S and SE England from Africa, Apr–Sep. More widespread during migration. Prefers deciduous open woodland, woodland edges and dense mature scrub.

Pied Flycatcher *Ficedula hypoleuca* 12–13.5cm

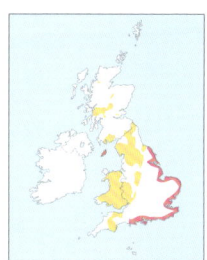

♀

Small, compact, with smart plumage. Breeding male black above, white wing-patch and small white forehead-patch. Underparts pure white. Adult female, non-breeding male and first-winter show brown to upperparts, with variable dark brown-black wings and reduced white wing-patch, being an angled bar across and then down wing instead of solid patch. Head brown, no white forehead. Underparts buff-white. Juvenile scaly above and below. Call a sharp rising *whit* or *zzz*, song sweet, simple melodic rising and falling notes.

Where to see: Locally common summer visitor, widespread passage migrant, Apr–Sep, winters in Africa. Breeds mainly in W Britain. Prefers mature woodland, particularly western oak woods, readily uses nest boxes. Coastal habitats mainly along E and S coast on migration.

♂

Red-breasted Flycatcher *Ficedula parva* 11–12cm

Small, dumpy flycatcher, long wings, rounded features. Perches with wings down, tail up and regularly flicked. First-year and females brown-grey above, white below, warm buff through flanks. Plain head shows black eye and fine white eye-ring. Tail distinctive, dark brown-black, two large white rounded patches each side of uppertail, most obvious in flight or take-off. Adult male similar but head and nape grey, throat and upper breast a neat orange-red. Fine bill, short legs black. Juvenile has buffy bar on wing. Call short, hard *tut*.

Where to see: Scarce passage migrant from N and E Europe, most Sep–Oct, along E and S coast from Shetland to Scilly Isles. Rare in spring. Winters in India, Pakistan. Visits coastal gardens, woodland edges, sheltered quarries, cliffs and valleys.

1st win.

Redstart *Phoenicurus phoenicurus* 13–14.5cm

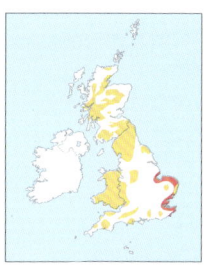

♀

Small, upright, slender chat, regularly quivers tail. Male shows grey back, nape and crown, brown-grey wings. Face solid black with contrasting white forehead. Underparts orange-brown, paler through belly. Male, female and first-winter show strong orange rump and tail, brown central feathers through tail. Female uniform mid-brown above, peachy-buff below, plain brown head. First-winter male greyer-headed, stronger orange underparts. Call sweet, rising *hooweet*, song soft, short warble with flourishing finale.

Where to see: Locally common summer visitor, uncommon passage migrant from Fennoscandia, Apr–Oct, winters in Africa. Mainly N and W England, Scotland and densest in Wales, patchy elsewhere. Prefers mature oak or mixed woodland, often elusive in canopy, nests in tree holes and nest boxes. More widespread along E coast during migration.

♂

Black Redstart *Phoenicurus ochruros* 13–14.5cm

♀

Smart, upright ground chat, distinct urban preferences. Breeding male distinct, smoky-grey throughout, including crown, black face and diffused black breast. Wings black with notable, variable white wing-panel. Vibrant orange rump and tail, latter with brown centre. Undertail pale orange-buff. Bill and legs black. Non-breeding male paler grey above and below, reduced black breast, browner wings. Female/first-winter like **Redstart** but colder, greyer plumage above and below, variable hint of pale wing-panel. Call low *tuk*, song short warble with dry crackling, contrasting sounds.

Where to see: Rare, local summer visitor from Europe, Mar–Sep, mainly SE and NW England. Scarce, widespread passage migrant and winter visitor through coastal UK, Sep–Mar. Prefers large, derelict, industrial buildings, coastal rocky shores in winter and on migration.

♂

Whinchat *Saxicola rubetra* 12–13cm

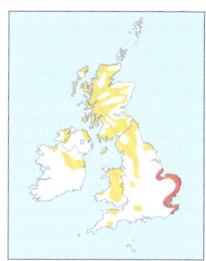

♂

Small, short-tailed, long-winged chat of open countryside. Male buff-brown and streaky above, head brown, prominent white supercilium and moustachial stripe. Underparts apricot-orange, strongest on throat and breast, paler towards belly and white undertail. Wing dark brown, white wing-panel. Female/first-winter similar, duller and paler throughout, buff supercilium, pale apricot-buff underparts, no white wing-panel. All show dark eyes, legs and brown tail with white base, obvious in flight. Call hard *tic*, song short, even-toned warble. Often perches on top of low vegetation or single plant stems.

Where to see: Scarce summer visitor and passage migrant from Fennoscandia, Apr–Oct, winters in Africa. Favours N and W upland and lowland heaths, moorland edges and rough grasslands. More widespread on S and E coasts during migration.

♀

Stonechat *Saxicola rubicola* 12–13cm

Short-winged, large-headed chat of open countryside. Male has full black-brown hood, partial white neck collar, dark brown upperparts, brown wings with white wing-patch, white rump flecked brown-black. Upper breast and flanks burnt orange, fading paler towards belly and white undertail. Female/first-winter paler, browner, with brown-buff hood and upperparts, weak whitish collar, reduced white wing-panel, paler orange underparts, warm brown and black-streaked rump. Juvenile shows short, faint, pale supercilium. Call hard *chak-chak* or high *whee*, song fast-chattering warble, often in brief flight display.

Where to see: Fairly common, local resident, common passage migrant and winter visitor from W Europe, Sep–Mar. Prefers open countryside, heaths with gorse, scrub and moorland, more coastal on migration and in winter.

♂ non-br.

♀

♂ br.

Wheatear *Oenanthe oenanthe* 14–16.5cm

♀

Slender, upright ground chat, long dark legs and wings. Breeding male blue-grey on back, nape, crown, wings black-brown, face with neat black mask, white supercilium. Underparts white, some apricot to throat. Female similar but drabber, lacks black mask. Non-breeding adults taupe-brown above, wings dark with pale fringes. First-winter sandy brown above, darker brown cheeks and wings, buff below. All show black band and black central block on white tail and rump, forming conspicuous 'T' shape, most notable in flight.

Where to see: Locally common summer visitor and passage migrant from N Europe, Mar–Oct, winters in Africa. Mainly N and W Britain, S and E coasts on migration. Breeds on moorland, grazed uplands, rocky open habitats. Coastal grasslands and large lawns, beaches on migration.

♂ br.

juv.

Dipper *Cinclus cinclus* 17–20cm

juv.

Unique, rotund waterside bird, deep chest and belly, short wings, stout, upright tail. Adult chocolate brown throughout, clean white throat and breast-patch. Small bill and long legs both dark. Juvenile cooler brown with pale fringes above, dusky with dark mottling below. Call high, hard *zit*, song more complex, wavering warble. Often observed 'body-bobbing' along river fringes, very adept at diving, swimming, walking in fast-flowing water, looking for aquatic invertebrates. Flight straight, fast, low along water courses.

Where to see: Scarce, widespread resident, mainly Scotland, N, NW and SW England, and Ireland. Also rare passage migrant to NE Britain from N Europe (Black-bellied Dipper, *C. c. cinclus*) Oct–Mar. Bird of upland and lowland fast-flowing streams and rivers through open habitats, woodland and urban areas.

House Sparrow *Passer domesticus* 14–16cm

Familiar garden bird with stout bill, and short legs and wings. Breeding male has grey crown and cheeks, chestnut eye-to-nape patch, and black bib and lores. Upperparts brown with black-and-buff streaking to back and wings. Bold white wing-bar, grey rump and pale grey underparts. Non-breeding male has reduced black bib, duller plumage. Female/juvenile pale sandy brown, upperparts with brown-and-buff streaks. Head pale brown with diffused grey cap, fine brown eye-stripe and broad buff supercilium. Underparts pale buff. Call uplifting cheeps and chirps, song repetitive cheeping.

Where to see: Common, widespread resident. Found in small to large sociable

flocks around human habitation. Often uses buildings as colonial nest-sites. Likes towns, villages, farmland, grasslands, parks, gardens and hedgerows.

Tree Sparrow *Passer montanus* 12–14cm

Smaller, neater and slimmer than **House Sparrow**, with similar but unique head pattern. Sexes alike. Easily identified by chestnut-brown crown, small, neat black bib, and obvious large white cheek-patch with central black spot and white collar. Upperparts pale sandy brown, black-and-buff streaking, white wing-bar. Underparts pale grey-white, flanks buff. Juvenile duller, with diffused chestnut crown and reduced black cheek-spot. Call an upbeat *tschirp* or hard *tek* in flight, song like high-pitched House Sparrow.

Where to see: Scarce, local resident with patchy distribution, mostly central and E England and Scotland and coastal Ireland. Social, flocking species, less associated with human activity. Nests in natural holes. Found in open countryside, woodland, farmland, hedgerows, parks, orchards and gardens.

Dunnock *Prunella modularis*   13–14.5cm

Small dark brown bird, sparrow-like, but often solitary, secretive, tendency to flick wings and tail, creeps and hops along while ground feeding. Adult dark brown above, bold black streaking, small white wing-bar of broken dots on uniform short brown wings. Head, neck, underparts grey, dark flecked brown crown, brown cheek-patch. Flanks warm brown with heavy dark brown streaking, underparts paler grey towards belly. Fine bill and dark, eye red on adult, dull brown on juvenile. Legs pink-brown. Juvenile like adult but with heavier streaking throughout. Call thin, piping *tseep*, song flat, fast warble, often from exposed perch.

Where to see: Common, widespread resident and winter visitor from N Europe, Sep–Mar. Frequents hedgerows, scrub, parks, gardens, farmland edges and woodland.

Yellow Wagtail *Motacilla flava* 15–16cm

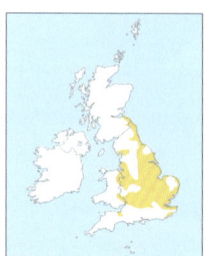

Smaller, shorter-tailed than other wagtails, slender, delicate profile. Adult male olive-greenish yellow above, brown-black white-edged wings and two notable narrow, pale wing-bars. Face and underparts striking lemon yellow, olive-green on crown and cheek-patch. Tail dark with white outer feathers. Bill fine and dark, legs thin, long and black. Female/juvenile similar, but duller olive-grey above, buff-yellow below, lack striking yellow tones. Juvenile browner above, buffer below with dark eye-stripe, cheek and necklace. Call high, piercing *tsree*. British breeding subspecies is *flavissima*; subspecies *flava*, *thunbergi*, *cinereocapilla* occur more rarely.

♀

Where to see: Scarce, local summer visitor, Apr–Oct, winters in Africa. Patchy distribution, mainly England. Prefers lowland open grazing marshes, wet grassland with livestock, arable fields, coastal and inland. More widespread during migration.

♂

Pied Wagtail *Motacilla alba*   16.5–19cm

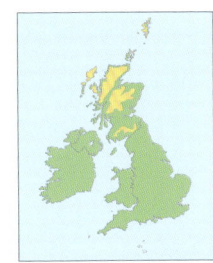

Common wagtail, long tail often 'pumped', smart black-and-white plumage. Adult male black above, white fringes to wing feathers, double white wing-bar. Head black, striking white facial disc-patch. Bib, upper breast black, underparts white with dark grey diffused flanks. Bill, eyes, long legs black. Tail black, white outer tail feathers, rump black/dark grey. Non-breeding female, juvenile similar but greyer on back and flanks, black bib and breast reduced to variable black breast-band. Call uplifting, high *chiz-ick*, often during undulating flight. Continental **White Wagtail** (ssp. *alba*) also occurs, adult grey-backed, contrasting with black nape, cleaner pale grey flanks, rump.

Where to see: Common, widespread resident, White uncommon passage migrant from Iceland, Greenland to continental Europe, rare breeder N Scotland. Open rural and urban areas.

Grey Wagtail *Motacilla cinerea* 17–20cm

juv.

Large, slender wagtail with long tail often characteristically pumped up and down. Adult male has storm grey upperparts, blackish-brown wing feathers, white-fringed tertial feathers. Head grey with black bib, white supercilium and sub-moustachial stripe. Underparts almost neon bright yellow around breast and undertail, flanks clean white, rump strong green-yellow. Adult female/non-breeding male similar but white bib with variable or no black. Juvenile has white bib, underparts whitish with pink-buff tinge, yellow on vent only. Call explosive *zi-zi*, sharper than **Pied Wagtail**.

Where to see: Locally common resident and winter visitor Sep–Mar, from N Europe. Prefers uplands summer, lowlands winter. Favours freshwater edges, streams, rivers, more widespread in winter, using marshes, lake fringes, gardens, water-filled ditches and roofs.

♀

Citrine Wagtail *Motacilla citreola*   15.5–17cm

Slimline wagtail, similar to **Yellow** and **Pied Wagtail**. Adult male has dark grey back, blackish-brown wings and tail, contrasting white feather fringes through wings and outer tail, two white wing-bars. Striking, unmarked bright yellow head bordered by narrow black shawl and yellow underparts. Female similar, but head with grey-brown crown and cheek-patch. First-winter grey, black and white, white feather fringes broader and wing-bar bolder than Yellow or Pied, face shows pale lores and grey cheek-patch encircled white, clearly separating cheek-patch from grey nape. Undertail very white, breast unmarked. Call distinct, rasping, high *tsriip*.

Where to see: Scarce annual migrant from Asia, mostly first-year birds in autumn, Sep–Oct, mainly S and E coast, from Scilly Isles to Shetland. Favours wet grasslands, pastures, marsh and wetland fringes.

♂ sum.

1st win.

Richard's Pipit *Anthus richardi*   17–20cm

Large pipit with upright stance, stout legs and strong, thrush-like bill. Close views show very long hind claw. Upperparts pale buff-brown, warmer orange tones to wing feather fringes, crown, ear coverts. Pale, wide supercilium, subtly marked pale lores, unlike dark-lores of **Tawny Pipit**. Underparts whitish with orange flank wash, dark streaking to upper breast only. Call distinctive, loud, rasping *schreep*, like House Sparrow. Flight strong, bounding.

Where to see: Scarce, widespread annual migrant from Asia, mostly Sep–Oct, E and SW Coast. Favours rough grassland.

Meadow Pipit *Anthus pratensis*   14–15.5cm

Small, streaky brown and buff, slender profile. Walks and feeds on ground, exposed or hidden in low vegetation. Warm olive-brown above, fine dark streaks, dark wings show buff feather fringes. Pale buff-yellow below, dense dark streaks to breast and flanks, cleaner whiter buff towards belly and undertail. Face plain, fine buff eye-ring, hint of buff eye-stripe. Dark tail shows white outer tail feathers. Bill dark, pink-based, legs pale, pink-orange.

Hindclaw long. Juvenile similar, richer toned. Call triple, high *sweet-sweet-sweet*. Aerial parachute display flight, often with *seep'seep* notes building into trill song.

Where to see: Common, widespread resident, passage migrant and winter visitor from N Europe. Breeds mainly in uplands, open countryside, moors, heaths, in winter, open lowlands, grasslands, coastal marshes and farmland.

Tree Pipit *Anthus trivialis* 14–16cm

Small, similar to **Meadow Pipit**. Olive-brown above, black streaked back, brown wings with pale buff feather fringes. Underparts whitish, buff-yellow with brown streaks to breast, finely streaked flanks, unlike thickly streaked Meadow Pipit. Hindclaw relatively short, legs pink. Readily perches in trees, alone or in pairs. Call buzzy, strong *tzee*. Song varied, long rise and fall notes in parachute song-flight.

Where to see: Scarce, widespread summer visitor, Apr–Oct, widespread passage migrant from N Europe, winters in Africa. Mainly Scotland, Wales, E and S England. Woodland edges, plantations, heaths, coniferous belts.

Water Pipit *Anthus spinoletta* 15.5–17cm

Large, pale, shy, most like **Rock Pipit**. Adult summer has grey-blue head, strong white supercilium and throat. Upperparts brown, faintly streaked, brown wings with buff feather fringes, dull white wing-bars. Underparts whitish, with variable pink wash, limited fine flank streaks. Bill and legs dark. Adult winter pale brown head and upperparts, buff-white below. Call loud, forced *weest*.

Where to see: Scarce, local winter visitor from S, C Europe, Sep–Apr. S, E bias in Scotland, Ireland, Wales and England, also C England. Coastal marshes, wetland including freshwater fringes, lagoons and estuaries.

Rock Pipit *Anthus petrosus* 15.5–17cm

Large, robust pipit, with cool, dusky tones, stout, relatively long dark bill, and dark legs. Adult grey-brown above with diffused dark streaking on back, wings grey-brown with dirty buff paler fringes and wing-bar. Head grey-brown, pale eye-ring, short, buffy, diffused supercilium, plain buffy throat. Underparts very dusky-grey and yellow-buff with smudged dark streaking throughout. Outer tail feathers grey not white. Note northern subspecies *littoralis*

appears in winter from Scandinavia in varying numbers. These show paler, cleaner underparts and more evident pale supercilium. Call loud, abruptly slurred *sveep*, song like **Meadow Pipit**'s, rising parachute songflight.

Where to see: Locally common coastal resident and winter visitor from N Europe, Sep–Apr. Frequents rocky shores, tideline, saltmarshes, small islands and cliffs; all coastal areas in winter.

Chaffinch *Fringilla coelebs* 14–16cm

Robust finch, long wings and tail, short, thick bill. Adult male has rich brown back and grey-and-white shoulder-patch. Wings blackish, pale yellow-green feather fringes and white wing-bars. Greenish rump in flight. Head shows blue-grey crown and nape, pink-orange face, silver-grey bill. Underparts uniform pink-orange, undertail white, tail black, legs pink-brown. Adult female/juvenile dull-toned, same wing pattern and rump as male, but back grey-brown, head pale grey with brown crown-stripe to side, and underparts uniform pale grey or variable pale brown. Call lively *pink* or plaintive, repetitive *huitt*, song fast, rising chatter with flourish finale.

♀

Where to see: Common, widespread resident and winter visitor from N Europe. Varied habitats with trees, woodlands, parks, gardens, farmland, rural and urban areas.

♂

Brambling *Fringilla montifringilla* 14–16cm

♀ non-br.

♂ non-br.

Winter visitor, similar profile as **Chaffinch**, often associates in ground-feeding flocks with them. Adult non-breeding male has blackish upperparts, extensive warm buff to orange feather fringes, looks scaly-backed with obvious orange shoulder-patch and double whitish wing-bar. Head mottled black-grey, neck pale grey, chunky bill yellow with black tip. Underparts pale orange through breast and flanks, white belly and variable black hindflank spots. Breeding male has full black hood, darker bill. Non-breeding female/first-winter similar to male but duller, paler, brown-grey head, buff-orange breast and dusky spotted peachy hindflanks. All show white rump in flight. Call nasal *tchway*.

Where to see: Uncommon, widespread winter visitor from N and W Europe, Sep–Apr. Lowland farmland, parks, gardens and woodland; favours beech tree mast.

Finches

Hawfinch *Coccothraustes coccothraustes*   17–18cm

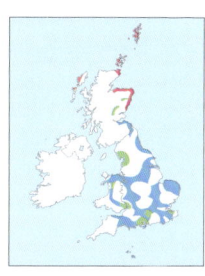

Stocky, distinctive, shy finch, large head, thick neck, very chunky bill, short tail. Male dark brown on back, grey neck collar, orange-brown head, rump and tail. Lores and bib black, bill conical, steel grey or pinkish. Wings show white wing-panel, glossy navy blue inner primaries. Underparts dusky pink-buff, white undertail, white tail-tip. Female similar, duller, with grey wing-panel. Juvenile pale-headed, lacks black on face, shows some dark flank barring. In flight, all front-heavy, obvious white in wings, tail. Call loud, hard *tick*.

Where to see: Scarce, local resident, rare passage migrant from N Europe. Breeding

♀

patchy, mainly Wales, England, more widespread in winter. Mature woodlands, parks, gardens and tree canopy of hornbeam, maple, beech and fruit trees; also readily ground feeds.

♂

Bullfinch *Pyrrhula pyrrhula* 15.5–17cm

♀

Rotund finch, bull neck, chubby black bill. Adult male shows grey back, black wings and tail, contrasting broad white wing-bar, neat black cap extending around bill. Face and underparts striking soft pink-red, undertail white. Female similar but grey-brown back, pale dusky pink face and underparts. Juvenile like female but plain grey-brown head. All show large white rump in flight. Call mournful, hollow, piping *peu*; song a quiet, soft whistle.

♂

Where to see: Uncommon, widespread resident and passage migrant from N Europe. Mainly S England and Ireland. More widespread in winter. Rare Northern Bullfinch (ssp. *pyrrhula*) occurs irregularly, mainly NE Scotland. Call hollow bugle, male striking rich pink-red. Found in woodlands, scrub, hedgerows, orchards and gardens. Often paired or in small groups.

juv.

Common Rosefinch *Carpodacus erythrina*   13.5–15cm

Stocky, charismatic, sparrow-sized finch, chunky head and bill, long wings and tail. Adult male has bright crimson-rose-red head, upper breast and rump. Wings and tail mid-brown, pale red-buff fringes and obvious white tertial fringes. Underparts whitish, flushed pink. Bill grey. Adult female olive-green-grey above, faintly streaked back, brown wings with pale feather edges and double wing-bar. Underparts white, olive flush, faint streaking. Juvenile similar to female, paler, streakier, stark beady black eye on plain face, softer, brighter upperparts. Wings show strong white tertial edging and wing-bar. Call and song soft, uplifting whistle notes and phrases.

Where to see: Scarce annual migrant from NE Europe, irregular rare breeder, May–Oct, winters in Asia. Mostly first-winter birds along S and E coast England and E Scotland, mostly Northern Isles.

juv.

Greenfinch *Chloris chloris* 14–16cm

♀

Large, heavy, thickset finch with strong bill. Adult male moss green on back and shoulders, wings grey and black, vibrant yellow on outer edges of primaries and tail. Rump vibrant yellowish-green. Head plain grey-green. Underparts green-yellow with grey flanks and white undertail. Bill and legs pale pink. Female similar but duller, browner, with faint streaking to brownish-grey back and greyer underparts. Juvenile like female, but underparts whiter with fine grey streaks. Call twittering *chichichichichit* and short *jup* repeated, song long, musical, nasal and wheezing.

Where to see: Uncommon, widespread resident and winter visitor, Oct–Apr, from N Europe. Favours lowland woodland, parks, gardens and farmland. More widespread, roaming with mixed ground-feeding finch flocks in winter. Comes readily to feeders.

♂

Twite *Linaria flavirostris*   12.5–14cm

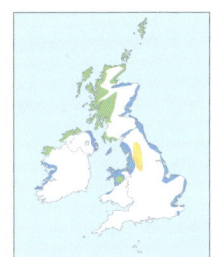

♀ win.

Small, buff-brown coastal finch, most like **Linnet**. Streaky brown-buff above, narrow whitish wing-bar and primary fringes. Small, finely streaked head, warm peachy-buff flush through face, throat. Underparts buff-white, soft streaking to breast, flanks. Breeding male has deep pink rump. All have small yellow bill in winter, greyish in summer, dark forked tail with white outer fringes, dark blackish eyes and legs. Call nasal, uplifting *twaay* or longer *tchwai*, song fast twittering, drawn-out nasal notes.

Where to see: Scarce, local resident, winter visitor Oct–Apr. Mainly coastal, moorlands in N and NW Britain. More widespread flocks along E coast Scotland, England in winter. Some upland birds disperse S in winter, birds from N Europe winter along E coast. Favours coastal rocky shores, cliffs, moorlands, farmland and saltmarshes.

♂ sum.

Linnet *Linaria cannabina*   12.5–14cm

Small, slender finch, longish forked tail, short grey bill. Breeding male striking, with strong pink-red forehead and breast, grey head and throat, rich chestnut upperparts, black-brown wings and tail with white feather fringes. Underparts white with warm brown wash, streaks to flanks. Non-breeding male duller, limited pink-red. Female/juvenile duller, cooler brown, softly streaked upperparts, greyish-brown head with small white cheek- and eyebrow-patch, finely streaked pale buff-white below. Call upbeat *tigg-it* or twangy *chi-chi-chi-chit*, song cheery, nasal mix of twittering trills.

Where to see: Uncommon, widespread resident, partial migrant and winter visitor, Sep–Mar. Northern birds disperse S in winter, some migrate to S Europe, others overwinter from N Europe. Open heaths, gorse, hedgerows, waste ground with scrub, more farmland and marshes in winter.

Common Redpoll *Acanthis flammea* 11.5–14cm

Like **Lesser Redpoll** but subtle discernible differences. Slightly larger, robust redpoll with colder grey-brown tones, lacking the warm brown-buff head and upperparts of Lesser. Whiter on wing feather fringes, back, rump and throughout underparts. Dark flank streaks starker on whiter plumage. Variable, can be tricky to separate in field. Some Common Redpolls are surprisingly buff, conversely Lesser can show grey-white tones, especially in winter. Call a hard *tjjit-tijit-tijit*.

Where to see: Uncommon, widespread winter visitor from N Europe, occasional breeder. Mainly along E coast in winter flocks, frequenting parklands, gardens and woodland.

Arctic Redpoll *Acanthis hornemanni* 12–14cm

Largest redpoll, bull-neck, cold grey-white frosty plumage, unstreaked large white rump. Some show buff to chunky head. Underparts white, encompassing rump, undertail-coverts, vent and upper-leg 'trousers'. Dark flank streaks stark on contrasting white plumage. May fluff up to appear 'snowball' like, showing extensive white. Back often cold grey-white with dark streaks, tail and wings almost black, with contrasting white wingbars.

Where to see: Rare, irregular, local visitor from N Europe, Oct–Mar, mostly N and E coasts, Scottish islands.

Lesser Redpoll *Acanthis cabaret*   11–14cm

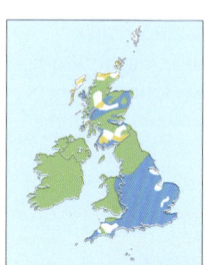

Small, dainty, streaky brown finch, tiny, pointed bill, forked tail. Upperparts buff-brown, darker brown streaking to back and rump. Tail and wings dark brown, paler outer edging and distinct buff-white wing-bar. Head shows buff wash, red cap, diffused black chin and bill surround. Eye dark, bill yellow with fine dark tip. Underparts white-buff with variable dark flank streaks. Breeding male has large red cap, crimson-pink-red to cheek, throat, breast and flanks. Female similar, buffier-faced and red only to smaller cap if present at all. First-winter buffier and streakier throughout. Short legs dark.

♂ br.

Where to see: Uncommon, widespread resident and winter visitor, mainly in N and W, and Ireland. Favours birch and alder woodland and conifer plantations; more widespread in winter, including gardens.

1st win.

Crossbill *Loxia curvirostra* 15–17cm

Very chunky finch, robust, thick-set body, large head, prominent bill. Upper and lower mandible tips curved and cross at tip. Rather plain though striking plumage, adult male showing uniform red-orange plumage throughout, brightest on rump with browner wings and tail, white lightly streaked undertail-coverts. Colour variation can see orange and green males as well as red. Bill and legs dark grey, tail notched. Female like male, but green throughout, often with some fine dark streaking. First-winter birds brown above, whitish below, heavily streaked throughout. Often in flocks, high up in conifer canopy. Call a *chip chip*.

Where to see: Scarce, widespread resident, irregular passage migrant from N Europe. Highly mobile, following favoured pine cone abundance. Prefers coniferous woodland.

♀

♂

Scottish Crossbill *Loxia scotica*   16–18cm

Slightly larger than **Crossbill**, with chunkier bill and lower, deeper *chup* call.

Where to see: Resident, confined to NE Scotland.

Parrot Crossbill *Loxia pytyopsittacus*   17–18cm

Rarest of the crossbills, large head, bull-neck and deep, chunky bill. Call a low, hard *tup*.

Where to see: Resident and passage migrant, breeds in conifer forests of NE Scotland. Sometimes occurs along E coast from Europe during irruption years. Identification in the field between all three crossbills can be very difficult, best identified from sound recordings.

Goldfinch *Carduelis carduelis* 12–13.5cm

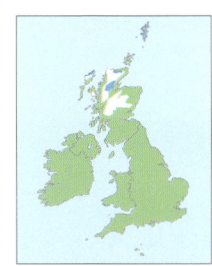

Small, delicate finch. Adult unmistakable, with red facial disc (extending past top of eye in male) surrounded by broad white band, black cap and nape collar. Bill pale grey-pink, pointed. Upperparts mushroom brown, wings and tail black with large white spots at feather tips, obvious golden-yellow wing-panel, very clear when perched or in flight. Underparts whitish, pale brown wash to flanks and breast. Juvenile has plain brown head, finely streaked back, breast and flanks. Call upbeat *tirilitt*, song twittering jangle.

Where to see: Fairly common, widespread resident, migrant and winter visitor Oct–Apr. Good numbers head S in winter, smaller numbers arrive from N and C Europe. Countryside, farmland, gardens and scrub, often on dandelions, teasel and thistle heads. Favours nyjer and sunflower seed feeders.

juv.

Serin *Serinus serinus* 11–12cm

Very small, streaked finch with a tiny, stout grey bill. All show yellow-green-brown upperparts with heavy dark streaks, wings dark with paler buff-green feather fringes, faint double wing-bar. Male has canary-yellow head, throat and breast, brown crown and cheeks. Underparts pale yellow to white with dark flank streaking. Female/juvenile similar, duller, heavy, dark streaking above and below, lack strong clean yellow seen in male. In flight, all show yellow rump. Call high-pitched, buzzing trill, song diagnostic, fast, high-pitched, almost electrical jingling.

Where to see: Very rare local resident and rare migrant from Europe. Breeding confided to S and SE coast of England. Most rare migrants, Apr–May, blown off course from continental Europe migration routes during southerly winds. Suburban woodlands, orchards and parks.

Siskin *Spinus spinus*   11–12.5cm

Small, short-tailed yellow-green finch, fine, pointed bill. Male striking, finely streaked olive-green back, black wings with yellow feather fringes, broad yellow wing-bar. Head yellow, black cap and bib. Underparts yellow, whiter towards belly, dark flank streaks. Female similar, duller, grey-green darkly streaked back and heavy streaking to whitish underparts. Face plain yellow-grey, blackish wings with yellow-white wing-bar. Juvenile like female but heavily streaked throughout.

All show yellow rump in flight. Call whistling two-note *teeyu-tsoooee*, song high trill.

Where to see: Common, widespread resident, winter visitor from N Europe. Breeds mainly in N regions, uncommon E Ireland, S and C England. Frequents coniferous and mixed woodlands and plantations. More widespread in winter, visiting parks, gardens, alder and birch trees. Readily visits feeders.

Lapland Bunting *Calcarius lapponicus* 16–17cm

♀ 1st win.

Large bunting, short legs, notably long wings. Low, elongated profile, bold patterning. Non-breeding male has warm buff head, dark crown and cheek edging, dark eye, pale pinkish bill. White moustachial stripe extends around cheek. Nape dark rufous. Back dark with broad pale-buff feather edges. Dark wing has rufous wing-panel. Underparts whitish with buffy, heavy, dark streaked flanks. Legs black. Non-breeding female/juvenile similar, paler-headed, dark cheeked with extensive dark streaking above. Breeding male striking, black head and throat, white extended eye-stripe, chestnut nape, yellow bill. Breeding female like non-breeding but with pale yellow bill and chestnut nape. Call a soft *teu*.

Where to see: Scarce, local winter visitor from Greenland, Fennoscandia, Sep–Apr. Rough coastal grassland and stubble fields along E coast.

♂

Snow Bunting *Plectrophenax nivalis* 15.5–18cm

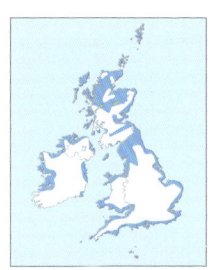

Long-winged, short-legged bunting, black, white and tawny-buff plumage. Breeding male white with black back, wings, central tail, and large white wing-patch. Bill, eye, legs dark. Breeding female similar but upperparts dull brown, pale whitish mottled feathering. Non-breeding adults show rich brown-yellow to crown and cheek-patch on white face, bulbous yellow bill. Back mix of black, white and buff mottling, wings black with extensive long white patch. Underparts white, some tawny-brown to breast, flanks. All show white-and-black wings and tail in flight. Call high, sweet, liquidy *tri-li-li-lip* often followed by a *tchew*.

Where to see: Uncommon, local winter visitor from Iceland, Fennoscandia, Sep–Apr. Rare breeder, Scottish Highlands. Lowland coastal sites in winter flocks, mainly shingle beaches, sand dunes and grasslands.

♂ non-br.

♀ non-br.

♂ br.

Corn Bunting *Emberiza calandra* 17–18cm

Large, robust, pale bunting, chunky bill. Upperparts pale, buff-clay brown, dark streaked back, dark brown wings with buff-clay feather fringes. Face rather plain, darker brown cheek-patch, finely streaked crown, dark eye with pale eye-ring. Throat and sub-moustachial stripe pale white-buff. Bill heavy, pale, straw yellow-pink. Underparts white-buff, bold, short dark streaks throughout, densest through upper breast with dark streaked central cluster. Legs pink. Call short *quit*, song distinct, short, rising though muted jangling trill, likened to shaking a bunch of keys. Male sings, bill open, head back, from low natural perches, wire fences, telegraph wires.

Where to see: Scarce local resident, very patchy distribution, mainly England, E Scotland, rare in Wales, Ireland. Favours open countryside, arable farmland, meadows and hedgerows. Forms small, often mixed flocks in winter.

Buntings

Yellowhammer *Emberiza citrinella* 16–17cm

Medium-sized bunting, longish tail, distinct coloration. Breeding male chestnut-brown above, heavy black streaking, chestnut-brown rump. Wings and tail with dark feathers edged chestnut brown. Head yellow, dark crown and cheek lines, neat grey bill. Underparts yellow, orange-brown streaked breast and flanks. Eyes dark, legs pinkish. Female/non-breeding male duller, less chestnut above, muted yellow, heavily streaked below. First-winter grey-brown above, buff-white below, very streaked. All show chestnut rump, white outer tail feathers in flight. Call hard *tzit* and liquid *pitilip*, song short rattle, likened to 'little bit of bread and no cheeeeese'.

Where to see: Locally common, widespread resident, winter visitor from Fennoscandia, Sep–Mar. Frequents farmland edges, hedgerows, scrub, heathland fringes, lowland grassland, and stubble fields in winter flocks.

♂ non-br.

♂ 1st win.

♀ non-br.

Cirl Bunting *Emberiza cirlus* 16–16.5cm

Slightly smaller, more compact than similar **Yellowhammer**. Breeding male chestnut brown, dark streaked back, chestnut-edged dark brown wings, bold head pattern, showing black eye-stripe and throat, yellow face, olive-grey crown, nape, upper breast-band. Underparts buff-white with yellow tones, chestnut along breast-sides, fine dark streaking to hindflanks. Rump olive-grey. Non-breeding male greyer through breast and nape. Female/juvenile buff-brown, duller, like female Yellowhammer, but olive-grey rump not chestnut, chestnut-toned shoulders, underparts more finely streaked, face with often obvious white cheek-spot. Call short, thin *tzit*, song rattling only *zezezeze*.

Where to see: Rare, very localised resident restricted to SW England, Devon, Cornwall. Found on farmland, especially meadows with old hedgerows near the coast. More arable fields in mixed feeding flocks in winter.

Ortolan Bunting *Emberiza hortulana* 15–16.5cm

Resembles **Yellowhammer**. Male has pale grey-green head, nape, neck, distinct white eye-ring, yellow throat and moustachial stripe. Bill and legs dark pink-red. Upperparts warm brown, with dark streaks, wing feathers dark with buff-chestnut fringes. Underparts pastel orange. Female duller, greyer head, subtle yellow to throat and moustachial stripe, paler orange below. First winter duller brown-buff, limited yellow-orange tones, heavier streaking below. All show cold brown rump. Call soft, liquid *plip*, metallic *plit* or *wink*.

Where to see: Rare migrant from Fennoscandia; winters in Africa. Most S and E coast, Sep–Oct. Grasslands, coastal scrub and stubble fields.

1st win.

Little Bunting *Emberiza pusilla* 12–13.5cm

1st win.

1st win.

Resembles female **Reed Bunting**. Bill small, grey, with straight upper mandible. Face shows white eye-ring and dark edged, chestnut lores and ear coverts, often with whitish spot at rear. Upperparts warm brown, with dark streaks. Whitish-buff below with heavy flank and breast streaks. Legs pink. Summer male has rufous central stripe on blackish crown.

Winter adult/first winter similar, duller, less rufous. Dark edging to hind-cheek only, stops short of bill base, unlike Reed Bunting. Call a sharp *tik*.

Where to see: Scarce, annual migrant from N Europe, mostly autumn, S and E coast, Sep–Mar. Coastal grasslands and scrub.

Reed Bunting *Emberiza schoeniclus* 14–16cm

♂ br.

Robust, slim wetland bunting. Breeding male warm brown above, dark streaks along back, chestnut-edged dark wing feathers. Head shows white neck collar and sub-moustachial stripe contrasting with solid black head, throat, upper central breast. Underparts white, grey wash to flanks, sparse dark streaking to hindflanks. In flight, white outer tail feathers on long dark tail and pale grey rump obvious. Winter male diffused brown-black head markings, streakier underparts. Female/ juvenile like non-breeding male, but pale throat, chestnut crown and eye-stripe, pale buff supercilium. Rump brown-grey, underparts heavily streaked. Call sharp, descending *tseeoo*, song repetitive, slow, short chips, trills, slurs. Often flicks tail.

Where to see: Fairly common, widespread resident, winter visitor from N Europe, Sep–Mar. Reedbeds, marshes, estuaries, riversides and farmland; open ground in winter.

♀

♂ non-br.

Glossary

Bib: a patch of colour covering the throat and upper breast

Blaze: contrastingly coloured marking around base of bill

Breast-band: a stripe of colour across the breast

Call: a simple sound made by a bird, for contact or to warn of danger

Crest: a tuft of feathers on top of a bird's head

Decurved: bending downwards

Drumming: repeatedly striking a resonant tree branch or trunk with the bill

Ear-coverts: upper cheek area, behind the eye

Ear-spot: a marking behind the eye

Ear-tufts: tufts of feathers on either side of the top of a bird's head

Eye-ring: a circle of coloured feathers or bare skin around the eye

Facial disc: the face of an owl, outlined by a ruff of short, stiff feathers

Fennoscandia: geographical region that includes Norway, Sweden and Finland. Also Kola peninsula and Karelia in NW Russia.

Flanks: sides of the lower body

Fringes: the edges of feathers, often contrastingly coloured to give a scaly appearance

Introduced: not native

Iridescent: having a brightly coloured sheen in certain lights

Leading edge (of wing): the front edge of the opened wing as the bird is flying

Lores: area between bill and eye

Mandible (upper/lower): the two parts of the bill

Mantle: the centre of the back

Pied: patched with black and white

Primaries: outermost, longest wing feathers

Secondaries: inner, shorter flight feathers in wing

Song: sound made by a bird (usually male) to advertise its territory; usually more complex than calls

Ssp: abbreviation of subspecies – a distinct form within a species

Supercilium: stripe above the eye, usually contrastingly pale

Tertials: the innermost secondary feathers

Trailing edge (of wing): the rear edge of the opened wing as a bird flies

Underparts: the bottom side of a bird – usually throat, breast, belly and flanks

Upperparts: the top side of a bird – usually the crown, neck, back, upperside of wings, rump and upperside of tail

Wattle: bare fleshy area on face, usually brightly coloured

Wing-bar: band of contrasting colour across wing

Acknowledgements

I would like to thank my parents Steve and Denise Nason and late grandparents Reginald (Jojo) and Betty Nason for a childhood filled with birds and travel. I hope my daughter Ayda grows to love and finds pleasure from birds and the natural world throughout her life as I have. Special thanks go to Jim Martin and Amy Hodkin at Bloomsbury for your invaluable contributions and assistance throughout and to copy-editor Lucy Beevor, designer Susan McIntyre, proofreader Marianne Taylor and indexer Angie Hipkin. Thanks also to Ian Andrews for your proofreading skills and to all the photographers who helped fill my species gaps with stunning imagery.

Photo credits

Bloomsbury Publishing would like to thank the following for providing photographs and for permission to reproduce copyright material within this book. While every effort has been made to trace and acknowledge all copyright holders, we would like to apologise for any errors or omissions, and invite readers to inform us so that corrections can be made to future editions.

Key to page positions: T = top; L = left; R = right; B = bottom; TL = top left; TR = top right; CL = centre left; C = centre; CR = centre right; BL = bottom left; BR = bottom right. **Abbreviated photo agency names:** AL = Alamy; Getty = Getty Images; RSPB = RSPB Images; SS = Shutterstock.

9 B Chris Gomersall; 10 B David Tipling; 13 T Erni/SS; 14 T Chris Gomersall; B Paul R. Sterry; 15 T CezaryKorkosz/SS; 16 T WildMedia/SS; 17 T Christopher Unsworth/SS; 18 T David Tipling; 19 T Fotogenix/SS; 20 B David Tipling; 22 T Henri_Lehtola/SS; 24 T David Tipling; B Erni/SS; 26 B David Tipling; 27 B Cliff Day/SS; 29 T David Tipling; 34 T 2009fotofriends/SS; 36 TR Brian Lasenby/SS; CR FloridaStock/SS; BL Feng Yu/SS; BR David Tipling; 41 T Erni/SS; B Vincent Legrand/AGAMI Photo Agency/AL; 42 T Grigorii Pisotsckii/SS; B David Tipling; 45 T Erni/SS; 48 T David Tipling; B David Tipling; 49 B Sokolov Alexey/SS; 50 T David Osborn/SS; B David Tipling; 51 T Dennis Jacobsen/SS; 53 B Rudmer Zwerver/SS; 54 T Erni/SS; 55 T Bob Gibbons/AL; B David Tipling; 56 T Dennis Jacobsen/SS; 57 T Alex Cooper Photography/SS; B David Tipling; 58 B David Tipling; 63 C Agami Photo Agency/SS; B David Tipling; 65 T Erni/SS; B David Tipling; 66 T David Tipling; B David Tipling; 71 T Maximillian cabinet/SS; 72 CL Agami Photo Agency/SS; CR plains wanderer/SS; B Chris Gomersall; 75 T David Tipling; 76 T P_vaida/SS; 78 T Tom Ingram/AL; C Erni/SS; 81 T David Tipling; 82 T David Tipling; CL David Osborn/SS; BL Voodison328/SS; BR David Tipling; 84 T Vincenzo Iacovoni/SS; 87 B FotoRequest/SS; 88 B Robert Schneider/SS; 93 T Gary Bell; B DaniloDjekovic/SS; 94 T David Tipling; 95 C Traveller MG/SS; 96 B David Tipling; 98 CR Sandra Standbridge/SS; 99 T Agami Photo Agency/SS; C Agami Photo Agency/SS; B Agami Photo Agency/SS; 101 T David Tipling; B David Tipling; 102 T RazvanZinica/SS; CL Traveller MG/SS; 105 B David Tipling; 108 T Dmitriy Drozd/SS; B David Tipling; 109 T HERGON/SS; CL David Tipling; CR David Jalda/SS; 110 T Sandra Standbridge/SS; 111 CL A Zargar/SS; CR moosehenderson/SS; B Dennis W Donohue/SS; 112 B Gregg Darling/Getty; 113 C Erni/SS; 115 B Traveller MG/SS; 116 B Agami Photo Agency/SS; 118 B David Parnaby; 123 T David Tipling; 124 T nexusby/SS; B David Tipling; 126 B Agami Photo Agency/SS; 128 T Chris Gomersall; B Chris Gomersall; 132 T Rory Tallack; 133 T David Tipling; C Chris Gomersall; 136 T DKeith/SS; B Taniaaraujo/SS; 137 T Maciej Olszewski/SS; 138 T David Tipling; B David Tipling; 139 T David Tipling; B Ivan Martin Hartono/SS; 140 T AlekseyKarpenko/SS; 141 T Peter Schwarz/SS; B Martin Prochazkacz/SS; 142 T LABETAA Andre/SS; B David Tipling; 143 T Michele Aldeghi/SS; C RMMPPhotography/SS; B scott mirror/SS; 145 T Martin Mecnarowski/SS; C HillebrandBreuker/SS; B Albert Beukhof/SS; 146 T Menno Schaefer/SS; B Chris Gomersall; 147 T Dimitar Rusev/SS; B SanderMeertinsPhotography/SS; 148 T Jesus Giraldo Gutierrez/SS; C Vitaly Ilyasov/SS; 149 T David Tipling; B davemhuntphotography/SS; 150 L Chris Gomersall; R Chris Gomersall; 151 T Wild Art/SS; C Dennis Jacobsen/SS; 152 B David Tipling; 154 T Brydon Thomason; 155 T David Tipling; 156 T Ian Grainger/SS; B Chris Hill/SS; 158 B David Tipling; 162 B Piotr Krzeslak/SS; 164 B Rory Tallack; 165 T Viktor Busel/SS; 166 BL Massimiliano Paolino/SS; 169 T Ercan Uc/SS; B Golore1955/SS; 170 T David Tipling; 171 TR WildlifeWorld/SS; CR WildMedia/SS; BR Monika Surzin/SS; 172 T Werner Baumgarten/SS; 173 T Keith Pritchard/SS; 174 T David Tipling; B David Tipling; 175 T SanderMeertinsPhotography/SS; 177 T Maciej Olszewski/SS; B David Tipling; 178 T Nick Vorobey/SS; 181 T Osipova Oleksandra/SS; 183 T David Tipling; 184 B Henri_Lehtola/SS; 186 B Yuriy Balagula/SS; 188 T Ilycsin/SS; B Chris Gomersall; 190 T Bouke Atema/SS; B David Tipling; 192 B Gallinago_media/SS; 194 B David Tipling; 195 T Phil Harris; 197 T David Tipling; B John Navajo/SS; 199 T David Tipling; B David Tipling; 201 T David Tipling; B David Tipling; 202 T Alex Penn; B Rory Tallack; 205 T Nick Vorobey/SS; B David Tipling; 206 B Marcin Perkowski/SS; 208 T David Tipling; 209 imageBROKER/Rainer Mueller/AL; B Agami Photo Agency/SS; 210 TL Agami Photo Agency/SS; TR John Navajo/SS; 211 B Rory Tallack; 218 T David Tipling; B Erni/SS; 225 T Paul Maguire/SS; 227 B Chris Gomersall; 229 T Ko Thongtawat/SS; 231 T stmilan/SS; B David Tipling; 232 B Keith Hider/SS; 233 BR David Tipling; 235 T Catleesi/SS; B David Tipling; 237 T Erni/SS; B Soru Epotok/SS; 238 T WaceQ/SS; 239 T Erni/SS; 241 T Mark robert paton/SS; 242 T David Tipling; 242 B Anna Shkuratova/SS; 244 T gergosz/SS; B David Tipling; 248 T Nick Vorobey/SS; 249 B Karin Jaehne/SS; 250 T Alex Penn; C Nick Vorobey/SS; 253 T Erni/SS; B David Tipling; 254 T Rob Christiaans/SS; 255 T Keith K/SS; 257 TL Max Carstairs/AL; TR Danny Green/RSPB; CR Johnny Madsen/AL; BR Dennis Jacobsen/SS; 258 BR Andy Wasley/SS; 259 T Babis & Sakis Tsilianidis; B David Tipling; 263 B Toni Genes/SS; 265 T Chris Gomersall; B Chris Gomersall; 267 BL Voodison328/SS.

Index

Acanthis cabaret 255
    flammea 254
    hornemanni 254
Accipiter gentilis 145
    nisus 144
Acrocephalus palustris
    201
        schoenobaenus 199
        scirpaceus 200
Actitis hypoleucos 93
Aegithalos caudatus 187
Aix galericulata 28
Alauda arvensis 189
Alca torda 120
Alcedo atthis 163
Alectoris rufa 13
Alle alle 118
Alopochen aegyptiaca
    26
Anas acuta 34
    carolinensis 36
    crecca 35
    platyrhynchos 33
Anser albifrons 22
    anser 19
    brachyrhynchus 21
    fabalis 20
    serrirostris 20
Anthus petrosus 245
    pratensis 243
    richardi 242
    spinoletta 244
    trivialis 244
Apus apus 49
Aquila chrysaetos 143
Ardea alba 139
    cinerea 137
    purpurea 138
Ardenna grisea 129
Arenaria interpres 80
Asio flammeus 161
    otus 160
Athene noctua 159
Auk, Little 118
Avocet 69
Aythya collaris 36
    ferina 37
    fuligula 38
    marila 39

Bee-eater, European 164
Bittern 135

Blackbird 219
Blackcap 204
Bluethroat 226
Bombycilla garrulus 180
Botaurus stellaris 135
Brambling 247
Branta bernicla 17
    canadensis 18
    leucopsis 18
Bubulcus ibis 136
Bucephala clangula 44
Bullfinch 249
Bunting, Cirl 265
    Corn 263
    Lapland 261
    Little 266
    Ortolan 266
    Reed 267
    Snow 262
Burhinus oedicnemus 67
Bustard, Great 50
Buteo buteo 152
    lagopus 151
Buzzard, Common 152
    Rough-legged 151

Calcarius lapponicus
    261
Calidris alba 88
    alpina 86
    canutus 81
    ferruginea 83
    maritima 87
    melanotos 87
    minuta 85
    pugnax 82
    temminckii 84
Capercaillie 9
Caprimulgus europaeus
    48
Carduelis carduelis 258
Carpodacus erythrina
    250
Cecropis daurica 194
Cepphus grylle 121
Certhia familiaris 215
Cettia cetti 194
Chaffinch 246
Charadrius dubius 74
    hiaticula 73
    morinellus 75
Chiffchaff 198

Chlidonias niger 115
Chloris chloris 251
Chough 174
Chroicocephalus
    ridibundus 100
Ciconia ciconia 133
Cinclus cinclus 235
Circus aeruginosus 146
    cyaneus 147
    pygargus 148
Clangula hyemalis 43
Coccothraustes
    coccothraustes 248
Coloeus monedula 175
Columba livia 52
    oenas 53
    palumbus 54
Coot 60
Cormorant 132
Corncrake 58
Corvus corax 179
    cornix 178
    corone 177
    frugilegus 176
Coturnix coturnix 15
Crake, Spotted 58
Crane 61
Crex crex 58
Crossbill 256
    Parrot 257
    Scottish 257
Crow, Carrion 177
    Hooded 178
Cuckoo 51
Cuculus canorus 51
Curlew 77
Curruca cantillans 210
    communis 208
    curruca 207
    iberiae 210
    nisoria 206
    subalpina 210
    undata 209
Cyanistes caeruleus 184
Cygnus columbianus 24
    cygnus 25
    olor 23

Delichon urbicum 193
Dendrocopos major 166
Dipper 235
Diver, Black-throated 124

    Great Northern 125
    Red-throated 123
Dotterel 75
Dove, Collared 56
    Rock 52
    Stock 53
    Turtle 55
Dryobates minor 165
Duck, Long-tailed 43
    Mandarin 28
    Ring-necked 36
    Tufted 38
Dunlin 86
Dunnock 238

Eagle, Golden 143
    White-tailed 150
Egret, Cattle 136
    Great White 139
    Little 140
Egretta garzetta 140
Eider 40
Emberiza calandra 263
    cirlus 265
    citrinella 264
    hortulana 266
    pusilla 266
    schoeniclus 267
Eremophila alpestris 190
Erithacus rubecula 225

Falco columbarius 154
    peregrinus 156
    subbuteo 155
    tinnunculus 153
Ficedula hypoleuca 228
    parva 229
Fieldfare 220
Firecrest 211
Flycatcher, Pied 228
    Red-breasted 229
    Spotted 224
Fratercula arctica 122
Fringilla coelebs 246
    montifringilla 247
Fulica atra 60
Fulmar 127
Fulmarus glacialis 127

Gadwall 31
Gallinago gallinago 90
Gallinula chloropus 59

Gannet 130
Garganey 29
Garrulus glandarius 172
Gavia arctica 124
   immer 125
   stellata 123
Godwit, Bar-tailed 78
   Black-tailed 79
Goldcrest 212
Goldeneye 44
Goldfinch 258
Goosander 46
Goose, Barnacle 18
   Brent 17
   Canada 18
   Egyptian 26
   Greylag 19
   Pink-footed 21
   Taiga Bean 20
   Tundra Bean 20
   White-fronted 22
Goshawk 145
Grebe, Black-necked 66
   Great Crested 64
   Little 62
   Red-necked 63
   Slavonian 65
Greenfinch 251
Greenshank 97
Grouse, Black 10
   Red 12
Grus grus 61
Guillemot, Black 121
   Common 119
Gull, Black-headed 100
   Caspian 108
   Common 103
   Glaucous 105
   Great Black-backed
   104
   Herring 107
   Iceland 106
   Lesser Black-backed
   110
   Little 101
   Mediterranean 102
   Sabine's 99
   Yellow-legged 109
Gulosus aristotelis 131

Haematopus ostralegus
68
Haliaeetus albicilla 150

Harrier, Hen 147
   Marsh 146
   Montagu's 148
Hawfinch 248
Heron, Grey 137
   Purple 138
Himantopus himantopus
67
Hippolais icterina 202
   polyglotta 202
Hirundo rustica 192
Hobby 155
Honey-buzzard 142
Hoopoe 162
Hydrobates pelagicus
126
Hydrocoloeus minutus
101

Ibis, Glossy 133
Ichthyaetus
   melanocephalus 102

Jackdaw 175
Jay 172
Jynx torquilla 164

Kestrel 153
Kingfisher 163
Kite, Red 149
Kittiwake 98
Knot 81

Lagopus lagopus 12
   muta 11
Lanius collurio 169
   excubitor 170
   senator 171
Lapwing 70
Lark, Shore 190
Larus argentatus 107
   cachinnans 108
   canus 103
   fuscus 110
   glaucoides 106
   hyperboreus 105
   marinus 104
   michahellis 109
Limosa lapponica 78
   limosa 79
Linaria cannabina 253
   flavirostris 252
Linnet 253

Locustella luscinioides
203
   naevia 203
Lophophanes cristatus
182
Loxia curvirostra 256
   pytyopsittacus 257
   scotica 257
Lullula arborea 188
Luscinia megarhynchos
227
   svecica 226
Lymnocryptes minimus
90
Lyrurus tetrix 10

Magpie 173
Mallard 33
Mareca penelope 32
   strepera 31
Martin, House 193
   Sand 191
Melanitta fusca 41
   nigra 42
Merganser, Red-breasted
47
Mergellus albellus 45
Mergus merganser 46
   serrator 47
Merlin 154
Merops apiaster 164
Milvus milvus 149
Moorhen 59
Morus bassanus 130
Motacilla alba 240
   cinerea 241
   citreola 242
   flava 239
Muscicapa striata 224

Nightingale 227
Nightjar 48
Numenius arquata 77
   phaeopus 76
Nuthatch 214

Oceanodroma
   leucorhoa 126
Oenanthe oenanthe 234
Oriole, Golden 171
Oriolus oriolus 171
Osprey 141
Otis tarda 50

Ouzel, Ring 218
Owl, Barn 157
   Little 159
   Long-eared 160
   Short-eared 161
   Tawny 158
Oystercatcher 68

Pandion haliaetus 141
Panurus biarmicus 186
Parakeet, Ring-necked
168
Partridge, Grey 14
   Red-legged 13
Parus major 185
Passer domesticus 236
   montanus 237
Pastor roseus 216
Perdix perdix 14
Peregrine 156
Periparus ater 181
Pernis apivorus 142
Petrel, Leach's 126
   Storm 126
Phalacrocorax carbo 132
Phalarope, Grey 92
   Red-necked 91
Phalaropus fulicarius 92
   lobatus 91
Phasianus colchicus 16
Pheasant 16
Phoenicurus ochruros 231
   phoenicurus 230
Phylloscopus collybita
198
   humei 195
   inornatus 196
   proregulus 196
   sibilatrix 195
   trochilus 197
Pica pica 173
Picus viridis 167
Pigeon, Feral 52
Pintail 34
Pipit, Meadow 243
   Richard's 242
   Rock 245
   Tree 244
   Water 244
Platalea leucorodia 134
Plectrophenax nivalis 262
Plegadis falcinellus 133
Plover, Golden 71

Grey 72
Little-ringed 74
Ringed 73
Pluvialis apricaria 71
squatarola 72
Pochard 37
Podiceps auritus 65
cristatus 64
grisegena 63
nigricollis 66
Poecile montanus 183
palustris 183
Porzana porzana 58
Prunella modularis 238
Psittacula krameri 168
Ptarmigan 11
Puffin 122
Puffinus puffinus 128
Pyrrhocorax pyrrhocorax 174
Pyrrhula pyrrhula 249

Quail 15

Rail, Water 57
Rallus aquaticus 57
Raven 179
Razorbill 120
Recurvirostra avosetta 69
Redpoll, Arctic 254
Common 254
Lesser 255
Redshank 95
Spotted 96
Redstart 230
Black 231
Redwing 221
Regulus ignicapilla 211
regulus 212
Riparia riparia 191
Rissa tridactyla 98
Robin 225
Rook 176
Rosefinch, Common 250
Ruff 82

Sanderling 88
Sandpiper, Common 93
Curlew 83
Green 94
Pectoral 87
Purple 87

Wood 94
Saxicola rubetra 232
rubicola 233
Scaup 39
Scolopax rusticola 89
Scoter, Common 42
Velvet 41
Serin 259
Serinus serinus 259
Shag 131
Shearwater, Manx 128
Sooty 129
Shelduck 27
Shoveler 30
Shrike, Great Grey 170
Red-backed 169
Woodchat 171
Siskin 260
Sitta europaea 214
Skua, Arctic 117
Great 116
Long-tailed 118
Pomarine 116
Skylark 189
Smew 45
Snipe 90
Jack 90
Somateria mollissima 40
Sparrow, House 236
Tree 237
Sparrowhawk 144
Spatula clypeata 30
querquedula 29
Spinus spinus 260
Spoonbill 134
Starling 217
Rose-coloured 216
Stercorarius longicaudus 118
parasiticus 117
pomarinus 116
skua 116
Sterna dougallii 112
hirundo 113
paradisaea 114
Sternula albifrons 112
Stilt, Black-winged 67
Stint, Little 85
Temminck's 84
Stonechat 233
Stone-curlew 67
Stork, White 133

Streptopelia decaocto 56
turtur 55
Strix aluco 158
Sturnus vulgaris 217
Swallow 192
Red-rumped 194
Swan, Bewick's 24
Mute 23
Whooper 25
Swift 49
Alpine 49
Sylvia atricapilla 204
borin 205

Tachybaptus ruficollis 62
Tachymarptis melba 49
Tadorna tadorna 27
Teal 35
Green-winged 36
Tern, Arctic 114
Black 115
Common 113
Little 112
Roseate 112
Sandwich 111
Tetrao urogallus 9
Thalasseus sandvicensis 111
Thrush, Mistle 223
Song 222
Tit, Bearded 186
Blue 184
Coal 181
Crested 182
Great 185
Long-tailed 187
Marsh 183
Willow 183
Treecreeper 215
Tringa erythropus 96
glareola 94
nebularia 97
ochropus 94
totanus 95
Troglodytes troglodytes 213
Turdus iliacus 221
merula 219
philomelos 222
pilaris 220
torquatus 218
viscivorus 223
Turnstone 80

Twite 252
Tyto alba 157

Upupa epops 162
Uria aalge 119

Vanellus vanellus 70

Wagtail, Citrine 242
Grey 241
Pied 240
Yellow 239
Warbler, Barred 206
Cetti's 194
Dartford 209
Eastern Subalpine 210
Garden 205
Grasshopper 203
Hume's 195
Icterine 202
Marsh 201
Melodious 202
Moltoni's 210
Pallas's 196
Reed 200
Savi's 203
Sedge 199
Western Subalpine 210
Willow 197
Wood 195
Yellow-browed 196
Waxwing 180
Wheatear 234
Whimbrel 76
Whinchat 232
Whitethroat 208
Lesser 207
Wigeon 32
Woodcock 89
Woodlark 188
Woodpecker, Great Spotted 166
Green 167
Lesser Spotted 165
Woodpigeon 54
Wren 213
Wryneck 164

Xema sabini 99

Yellowhammer 264